Quantum Tongues

The Convergence of Physics and Linguistics

Siena Avery Shah

Quantum Tongues
Copyright ©2024 Siena Avery Shah
ISBN 978-1506-914-45-9 PBK
ISBN 978-1506-914-46-6 EBK

October 2024

Published and Distributed by
First Edition Design Publishing, Inc.
P.O. Box 17646, Sarasota, FL 34276-3217
www.firsteditiondesignpublishing.com

In physics, the truth is rarely perfectly clear, and that is certainly universally the case in human affairs. Hence, what is not surrounded by uncertainty cannot be the truth.

Richard P. Fenyman

Contents

Introduction

When it comes to interpreting the world as a whole, there are two approaches that, at first glance, seem to be irrevocably different. The physical approach, which, to put it simply, views the world as all of the processes that make it up, and the linguistic approach, which views the world as the people within it develop through the lens of their languages. While these approaches focus on two incredibly different, but important, aspects of the world, the processes of these approaches are more similar than they first appear to be. Both physics and linguistics focus on the structure of the world, whether that structure be the behavior of energy and matter or the languages that make up the world, and because they both focus on the structure of the world, they both rely on the interpretation of the complex patterns that emerge from the close study of their "worlds". Both approaches are ultimately focused on the transmission of information as well, as parts of physics are focused on the signals transported by particles, while parts of linguistics are involved in the meaning transported by the words and tone utilized by people. Most importantly, by combining these two different fields, we are able to look at a more complete view of the world, not just the physical world, and not just the human aspect of the world, in all its contradictory complexity.

Physics itself serves as the language of the universe by allowing us to comprehend the universe through observations and theories as a result of those observations. Mathematics, which is used as the language of physics, is used to describe the behavior of matter, energy, and particles in the universe and is used to express physical theories and observations. Because of the way they use mathematics, physicists can express complex physical phenomena the same way that humans use language every day in order to express complex thoughts and ideas.

The same way that the study of physics utilizes mathematics in order to express complex phenomena, the study of linguistics utilizes language in order to express complex thoughts and ideas. Through the usage of linguistics, we can see the development and complexity of semantics, syntax, tone, and structure of different languages depending on the culture the language developed within, and what the language was built for. For example, because ASL, or American Sign Language, was created for deaf and hard of hearing people, ASL as a whole is based on the concept of *showing*, rather than verbally *saying*. Direct objects, for example, are expressed first, then adjectives are signed. If someone were to say a statement about gray curtains, the speaker would first use the sign for curtains, allowing the listener to visualize the curtains that the color gray will be assigned to, rather than visualize the concept of something gray, that will then be curtains. Tone in ASL is also indicated based on the way that speakers act out the signs, adding nuance to the language that cannot be directly expressed verbally. Just like how tone in ASL cannot have a direct verbal translation in English, languages don't have identical ways of expressing certain concepts. Without the study of linguistics and the cultural nuance as a result of certain languages, or the ways that those languages developed because of those cultures, many cross-cultural miscommunications can happen, resulting in misunderstandings of different cultures, dehumanization of certain peoples, and atrocities as a result of one culture viewing another culture as inherently inferior. Linguistics emphasizes the crucial role of language in facilitating communication, and the necessity of understanding those who may not share a language similarly to how physics emphasizes the necessity of mathematics in expressing complex physical phenomena.

By utilizing the intersection between physics and linguistics, we open up a whole new world of possibilities for what we can discover about both the universe and ourselves. However, a key issue is currently plaguing the field of linguistics. According to UNESCO, around 3,000 languages have the potential to disappear by the end of the century, meaning languages will die out at a rate of roughly one language every two weeks. The death of so many languages will have many profoundly negative consequences, despite being a tragedy in and of itself, one of which being the "reduction" of the universe that linguistics can analyze. If so many languages die, the linguistic "universe" will drastically reduce in size, another tragedy. So, what exactly are the consequences of these deaths, why should we put such great effort into preserving these languages, and how can we go about doing so?

Chapter One
The Genesis of Language

Language, in its origin and essence, is simply a system of signs or symbols that denote real occurrences or their echo in the human soul.

Carl Jung

Universal Development of Language

When we compare the frequencies of words of early languages, we can gain valuable insights into the patterns/structures of language evolution. Despite the variations in vocabulary and grammar that may be present across different languages, linguistic research has shown some similarities in word frequency across these different languages. By examining the frequency of words in early languages, researchers can gain a better understanding of language development and the usage of certain words.

Because studies have demonstrated that certain words tend to occur with higher frequencies across multiple distinct languages, a degree of universality in how linguistic information is encoded is suggested. For instance, common concepts such as earth, water, and food are likely to be represented by similar sounding words in various languages. This indicates that certain linguistic categories may consistently influence word usage patterns, regardless of the specific language being analyzed.

Zipf's law describes a mathematical relationship between the rank of a word and its frequency in natural languages. After assigning a rank to words, with the first rank being the word that shows up the most, the second rank being the word that shows up the second most, and so on and so forth, the frequency is related to that rank by roughly $f(r) \propto \frac{1}{r}$. This means that the frequency of the rank two work occurs at roughly half the rate of the rank one word, the rank three word occurs at roughly a third of the rate of the rank one work, and so on. This relationship is quite puzzling, and occurs even in many man-made languages such as Esperanto. While not much is known about why this relationship occurs, one such explanation is related to the preferential attachment process; due to the success of the rank one word and other high-ranking words, they are more commonly used. Meanwhile, less common words are less likely to spread, making them be used less. Zipf's law shows us that the similarity between commonly used words, such as *food* or *water* may be partially due to how commonly they are used.

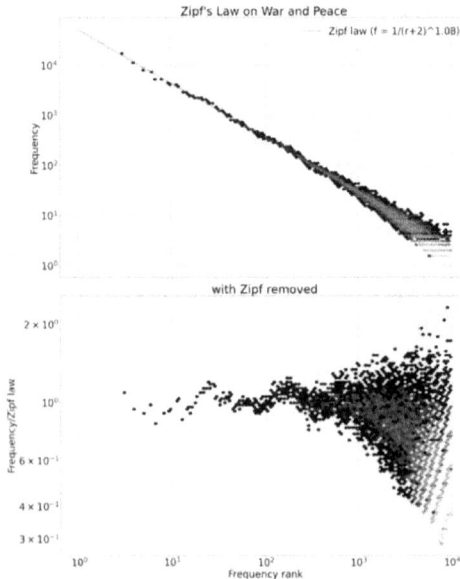

Zipfs law on war and peace leyaiuafhdglkajhd

Another aspect to consider is how cultural factors may influence word frequencies. Different cultural factors can shape the frequency of specific words within a given language system. For example, in a hunter-gatherer society, more words pertaining to hunter-gathering may be prevalent than words pertaining to farming, whereas in agricultural societies, more words relating to farming and agriculture may be common. Another example is in Inuktitut, the language of the Inuit people. Due to how common snow is where the Inuit people live, the Inuit have dozens of words for snow, while many aboriginal Australian languages do not have any words for snow.

Investigating how frequently words in early languages occur can provide a valuable lens through which to view language evolution and usage. By examining the distribution of word frequencies across different languages, researchers can discover important insights into the universality and cultural specificity of languages and how they relate to one another.

Certain phonetic patterns, referred to as "linguistic universals" have been observed to transcend language. There are several different types of linguistic universals, many of which can coexist. Absolute linguistic universals are statements that remain true for every language, such as the statement *all languages have vowels*. Implicational universals are statements that apply to languages that have a certain feature that must be accompanied by another feature, such as the statement *if a language has gender in nouns, it has gender in pronouns*. Tendencies are statements that are not true for every language, but that are true for too many languages to be coincidence.

There are two main possible reasons for the existence of linguistic universals. The first is the existence of universal grammar, or the idea that language is something that humans can innately acquire. Because of the theoretical innate ability to acquire language, many languages develop similar structures and patterns. Another reason for the existence of linguistic universals is simply because communication, the main facet through which all language hinges on, would be greatly hindered without the presence of a linguistic universal law. Therefore, even if a language was artificially created to lack a linguistic universal, eventually, it would gain that linguistic universal.

The word for "mother" contains bilabial sounds, or sounds made by pressing your lips together, across many languages. The *m*, *b*, or *p* sounds, often accompanied by an open vowel like *a*, are particularly prevalent in the word for "mother" across languages. This pattern can be seen in the following examples:

Indo-European Languages

English: *Mother*
Latin: *Mater*
Sanskrit: *Mātṛ*

Dravidian Languages

Tamil: *Amma*
Telugu: *Amma*

Semitic Languages:

Arabic: *Umm*
Hebrew: *Imma*

The usage of these sounds across different languages is hypothesized to be because of the natural ease with which babies make these sounds. The [m] sound is one of the first consonants that infants can sound out, often while nursing. The soft *a* sound is also one of the earliest vowels to be produced, making *ma* a likely candidate for early vocalization, resulting in that syllable being present in many words that mean mother.

Some linguists propose a phonosemantic motivation, or a direct connection between the way a word sounds and its actual meaning, behind these similarities. This suggests that the *ma* sound may have some inherent associations, such as softness, warmth, and nurturing, all things that are often associated with the concept of a mother.

This phenomenon is not limited to the word "mother." In fact, other basic terms, particularly those related to familial relationships, body parts, and onomatopoeias, often show similar cross-linguistic patterns. For instance:

Father: Often characterized by [p], [b], or [f] sounds.

- English: Father
- Latin: Pater
- Sanskrit: Pitṛ

Eye: Words for "eye" often contain [k] or [g] sounds.

- Greek: Ophthalmos
- Old Norse: Auga
- Sanskrit: Akṣi

Meow: Often characterized by [me] sounds

- Kazakh: мияу (mijau)
- Tagalog: Meyow
- Persian: Mioo

The recurrence of phonetic patterns across languages, especially in certain fundamental words such as "mother" or "father", demonstrates the intersection between human physiology, cognitive processes, and linguistic evolution. These universalities that occur across many different languages not only show us insights into the development of language but also highlight the deep-seated connections between varieties of different human cultures, regardless of physical distance between them.

While there are many differences across many different cultures that occur as a result of the unique climates, societies, and events that happen across those cultures that result in the different developments of their languages, we can see that there are still some universalities across these languages due to a variety of different processes and theories.

The integration of emojis into modern communication represents an intriguing evolution in language, blending visual elements with textual communication to convey meaning, emotion, and nuance in ways that traditional language might not easily capture. From a scientific perspective, particularly within the fields of physics, information theory, and cognitive science, the rise of emojis can be analyzed in terms of their impact on information transmission, their role in the evolution of language, and their potential to shape future forms of human communication, something incredibly important.

Emojis function as a kind of visual shorthand, capable of expressing emotions, concepts, or even complex ideas in a single symbol. This symbolic compression of information aligns closely with principles from information theory, particularly the concept of entropy in communication systems. In information theory, entropy measures the uncertainty or unpredictability of a message; a higher entropy implies more possible meanings or interpretations. Emojis, by condensing complex emotional or contextual cues into simple visual forms, reduce the entropy of a message by providing immediate, often universally recognized signals that clarify the intended meaning. This can be seen as an optimization of communication, where the redundancy and potential ambiguity of text are minimized by adding a layer of visual context.

The physics of visual perception also plays a crucial role in understanding how emojis impact communication. Human vision and cognition are highly attuned to recognizing and interpreting visual symbols rapidly and efficiently. Gestalt principles from visual perception, which describe how humans naturally group and interpret visual stimuli, can help explain the effectiveness of emojis. For example, emojis are often designed with simplicity and high contrast, allowing them to be easily distinguished and recognized even in complex or cluttered digital environments. This efficiency in recognition is similar to how the brain processes and recognizes speech patterns, and it suggests that emojis are particularly well-suited to complement or even enhance textual communication.

When observing through the lens of linguistic evolution, the emojis could theoretically be the next step towards a more universal language. Historically, written language has evolved from visual representations of ideas, and then into the complex symbols such as alphabets that we use today. Emojis might represent a reversal or re-emergence of these early forms of writing, where visual symbols carry significant semantic weight. In this sense, emojis could be viewed as a modern-day equivalent of ancient hieroglyphs or ideograms, offering a parallel evolution in digital communication.

Looking to the future, one could even envision the development of quantum communication systems that incorporate the use of emojis as part of their encoded information. Just as quantum bits (qubits) can represent and transmit information in ways that classical bits cannot, an advanced digital communication system might use emojis as a means of conveying information that is context-dependent and dynamically interpreted, enhancing the richness and efficiency of communication.

Moreover, the potential for emojis to influence the evolution of language extends beyond digital communication. As more people use emojis across different languages and cultures, these symbols may start to influence spoken and written language in more profound ways. For instance, certain emojis might become embedded within everyday speech as verbal expressions (e.g., saying "heart" instead of expressing love) or even evolve into new linguistic constructs that transcend language barriers, leading to a more hybrid form of communication that blends traditional language with visual symbols.

Tone and language

Spoken language uses physics in the form of wave acoustics and tone. Tone, which refers to the perceived pitch of a sound, is determined by the sound wave's frequency, produced during speech. Tone is a crucial element that conveys not only the basic phonetic properties of words but also meaning, such as emotional nuance, intent, and even syntactic or grammatical distinctions in tonal languages. Understanding the physics of tone involves analyzing the properties of sound waves, the resonance of the vocal tract, and the auditory perception of pitch.

Tone in language is produced by the vibration of the vocal cords in the larynx, at its most basic level. These vibrations are capable of creating periodic pressure waves that travel through the air as sound. The frequency of these vibrations determines the pitch of the sound, with frequencies and pitches being directly proportional. The fundamental frequency of the vocal cords, which will be expanded on later in this book, is the primary determinant of the perceived pitch of a speech sound. It is the final tone, however, that listeners perceive as shaped not only by this fundamental frequency, but also by the harmonics and formants that arise from the resonant properties of the vocal tract.

The vocal tract acts as a resonator by amplifying certain frequencies while dampening others. These amplified frequencies are known as formants, which play a critical role in shaping the acoustic characteristics of the sounds of speech. In tonal languages, the variation in the fundamental frequency over the course of a certain word, also known as the tone contour, carries a linguistic meaning. For example, in Mandarin Chinese, the word "ma" can have different meanings depending on its tone: it can mean "mother," "hemp," "horse," or "scold" depending on its tone.

Analyzing tone in spoken language involves using tools such as spectrograms and pitch trackers to visualize and measure the frequency components of speech over time. Spectrograms show how the energy of different frequency components evolve, therefore allowing researchers to identify the pitch, formants, and harmonics that make up the tone. Pitch trackers specifically measure the fundamental frequency of speech and plot how it changes over time, providing an in-depth picture of the tone contour.

For instance, a pitch tracker analyzing the word "ma" in Mandarin would show different pitch trajectories depending on the intended tone, with each trajectory corresponding to a distinct lexical meaning. By analyzing these pitch patterns, linguists can gain insights into the role of tone in communication and how it varies across different languages and dialects.

The physics of tone also plays a rather large role in how spoken language is perceived. The human auditory system is tuned to detect variations in pitch, allowing listeners to observe certain subtle differences in tone that may be linguistically significant. In fact, the specific differences in tone that can be observed varies based on the languages that a person speaks; a person who speaks Farsi, for example, would be able to detect nuances in tone that someone who exclusively speaks English or Mandarin Chinese would be unable to. This sensitivity to pitch is essential for understanding tonal languages, where misinterpreting a tone could result in some misunderstandings. However, even in non-tonal languages, tone can contribute to certain aspects of speech, thereby helping to convey attitudes, emotions, and emphasis.

The improvement in the clarity of spoken and written communication can benefit from a deeper understanding of the physics of tone. For example, speech synthesis and recognition technologies rely on accurate modeling of tone and pitch to produce natural-sounding speech and to accurately transcribe spoken language into text, things that can therefore be used in order to preserve language. By incorporating detailed acoustic models that account for the resonant properties of the vocal tract and the dynamics of pitch variation, these technologies can improve their accuracy and effectiveness.

The concept of tone can be extended metaphorically to represent the intonation patterns as well as the prosody that are often implied in text but that are unable to be explicitly represented. Prosody refers to the rhythm and stress of speech, which can together contribute to the overall tone of spoken language. For example, the prosody of a question in English is often what distinguishes a question from a statement. While written language lacks these auditory cues, certain syntactic conventions, such as punctuation, are able to be used to convey aspects of prosody. Unfortunately, these conventions are often incapable of fully capturing the richness of spoken intonation, leading to potential ambiguities in interpretation.

In order to address this challenge, researchers have explored the use of prosodic annotations and text-to-speech systems that are capable of incorporating detailed prosodic information. They do this by annotating text with prosodic markers that are capable of indicating pitch, stress, and rhythm; it is therefore possible to create written representations that more closely align with the intended spoken tone. When these annotations are used together with advanced text-to-speech systems, the resulting speech output can more accurately reflect the intended meaning and emotional content of the original text.

Another promising area of research is in regards to the use of neural networks as well as machine learning algorithms, which can be used in order to model the complex relationships between text and prosody/the intonation of language. These models can learn from large datasets of spoken language to predict the expected intonation of a given text, which can enable a more natural and expressive speech synthesis. By integrating these models into certain communication technologies, we can enhance the clarity as well as the expressiveness of spoken language, especially in connection to written language, which can often lack that expressiveness unique to spoken language, therefore making it easier for users to both convey as well as perceive certain subtle nuances in a text's tone.

More than that, tone in verbal language is multifaceted, and not only conveys the pitch of speech sounds, but also carries large amounts of significant information about the speaker's emotional state, intent, and can even provide crucial social context. Without getting ahead of ourselves, the physics of tone relates to the acoustic properties of sound; particularly the frequency and amplitude of vocal vibrations, which are both shaped by the producer's vocal tract and perceived by the listeners auditory abilities, which are directly related to the shape of their ear canals and other factors. However, tone extends beyond these physical properties to include the broader prosodic, and more subtle, features of speech, such as intonation, stress, rhythm, and voice quality. These features collectively contribute to the expression of different emotional states or attitudes, such as anger, happiness, or sarcasm.

In the context of conveying emotions or attitudes, tone is not just about the pitch in a certain moment (or the fundamental frequency, for that matter) but also about how that pitch changes over time, its volume, and the quality of the voice (such as being breathy, tense, or relaxed). All of these complex elements together combine to create a signal that listeners interpret to infer the speaker's emotional state, though, due to the ambiguity of emotional states and the ways that they are expressed based on culture or the individual, oftentimes it can be difficult to infer a speaker's emotional state.

Now, let's take a look at an example of how pitch and other prosodic elements relate to understanding a speaker's tone. An angry tone is typically characterized by a higher overall pitch, increased loudness, and a more tense/harsh voice quality. The pitch typically rises rapidly and can have sharp, abrupt changes, usually indicative of the speaker's heightened emotional state. Physically, this tension in the voice typically comes from increased muscle tension in the vocal cords and vocal tract, which alters the acoustic properties of the sound waves therefore produced. The spectrogram of an angry tone would likely show more energy in the higher frequency bands and a more irregular pitch contour, indicating the abrupt shifts in pitch that are often associated with anger.

In contrast, while a happy tone typically features a higher pitch, similarly to an angry tone, it usually occurs with smoother and more mellifluous pitch contours. The speech might be louder, similar to anger, but with a warmer and more resonant voice quality, indicating relaxation rather than tension. The transitions between pitch levels in happy speech are usually more fluid, with a rising and falling intonation that suggests positivity and openness. The acoustic energy is more evenly distributed across the frequency spectrum, and the pitch contours show gradual changes rather than the sharp, erratic shifts seen in angry tones.

While looking at specific cases of how certain tones are created and perceived, it is worth looking at sarcasm. Sarcasm often involves a deliberate manipulation of tone to convey a meaning opposite to the literal words spoken. A sarcastic tone often involves a flat or exaggerated pitch, where the speaker may use an intonation pattern that is overly smooth or monotone, contrary to what would be expected for the content being conveyed. For instance, a speaker might say "Great job!" with a falling intonation and a low pitch, which would normally suggest disappointment or criticism rather than praise. The voice quality might also be slightly nasal or drawling, adding to the impression that the speaker does not genuinely mean what they are saying. In this case, pitch contour and voice quality depict to the listener that the speaker's intended meaning is contradictory to the speaker's actual words, therefore illustrating to the listener that the speaker is sarcastic.

To analyze how tone can depict different emotions, researchers often use tools like spectrograms, pitch trackers, and acoustic analysis software. A spectrogram can show how the frequency content of speech changes over time, revealing the patterns of pitch, formants, and intensity that correspond to different tones. For instance, by comparing spectrograms of speech with different emotional tones, one might observe that angry speech has more energy in the higher frequencies and more rapid changes in pitch, while happy speech shows smoother, more harmonic patterns, which aligns with the physics of how these tones are produced. A pitch tracker, on the other hand, can plot the fundamental frequency over time, allowing researchers to study the intonation patterns associated with different emotions.

Voice quality, another important factor in the quality of tone, can also be analyzed in regards to its acoustic properties. A tense or harsh voice quality, often found in angry speech, can be characterized by a greater degree of jitter, or frequency variation, and shimmer, or amplitude variation, as well as increased spectral noise, or the distribution of the signals of noise across different frequencies. These acoustic features contribute to the perception of an aggressive or strained tone. Conversely, a breathy or soft voice quality, which might be used to convey affection or sadness, would show less spectral noise and a smoother waveform, indicative of relaxed vocal cord vibrations.

The way that these tonal variations are perceived by listeners involves complex auditory processing in the brain. Most auditory systems are carefully designed to detect subtle differences in pitch, loudness, and timbre, therefore allowing us to interpret both the emotional content of speech, as well as the actual meaning of what is being conveyed. The brain combines these cues with its context, such as the words spoken and the situation the listener finds themself in, in order to infer both the speaker's intent and emotional state. This process is critical for effective communication, since it helps listeners understand not only just what is being said but also how it is meant.

Understanding tonal variations can improve the accuracy of the detection of emotion as well as intent in voice recognition systems. This is especially important in applications that may eventually be automated, such as customer service. In this scenario, the system that would replace a physical human being would need to not only be able to gauge the user's emotional state, but would also need to be able to respond appropriately, and as a result meet the consumer's needs.

Tone in verbal language displays a large range of features such as pitch, loudness, and voice quality. Together, these features convey the speaker's emotional state, intent, and more. The physics of tone, which will be discussed later in this book, heavily involves the interactions of these features and how they interact with properties of the vocal tract. Different emotions, such as those previously discussed, are also characterized by their common distinct tonal patterns.

Ecological Pressures on the Development of Oral Language

Before we can understand the direct consequences of the mass deaths of languages, we must first understand languages themselves, including their origins. During the genesis of language, it is predicted that human communication began with simple vocalizations and gestures, similar to how modern primates communicate today. It is likely that these vocalizations and gestures eventually became more and more complex as society itself evolved, resulting in the need for increasingly complex language, eventually resulting in languages with certain grammatical structures, and even with writing systems. Each language, and each resulting dialect of those languages, adapted due to its environment and the purpose it needed to serve, and, eventually, further adapted due to interactions with other languages and cultures.

One such theory that shows how languages have adapted due to their surrounding environments is the acoustic adaptation hypothesis. The acoustic adaptation hypothesis suggests that vocalizations have evolved in order to maximize their transmission within their environments. In other words, the transmission of vocalizations of certain animals, and the transmission of languages, are most effective in the language in which these languages developed. This hypothesis proposes that languages have developed based on the humidity levels of certain areas, how dense the population was, the presence of predators/prey or competitors, etc.

When we take a look at just the climate languages developed in, for example, we are given some incredible insights. In more humid, rainforest-y areas, languages tend to have less consonants and more vowels, as consonants tend to be more difficult to transmit through dense foliage/rain, and hot weather can distort consonants to a greater extent than it does vowels. Similarly, tones are much less likely to develop in arid regions due to the level of precision needed to actually pronounce them. Because our vocal cords are less precise in arid regions, languages from arid regions are much less likely to have tones. This is why so many humid countries, such as China and Vietnam, have languages that rely very heavily on their tone (shí, shī, shǐ, shì, can mean food, poetry, beginning, and persimmon respectively in Mandarin, without including each word's homophones), while many more arid countries have more toneless languages (think of how tone in English is used to express emotion rather than meaning). While the causation of linguistic traits because of ecological pressures is difficult to study accurately, and hasn't been studied much until recently, these patterns are still worth noting.

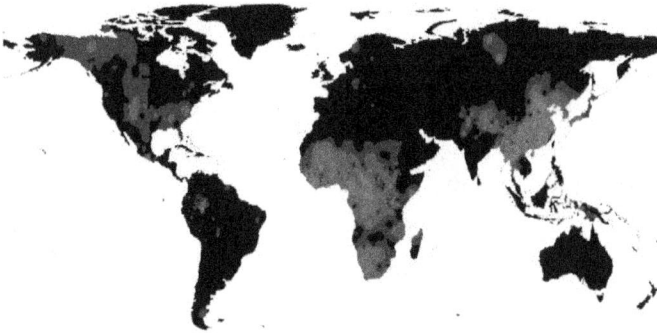

A heat map based on the density of languages with/without complex tonality (defined here as 3 or more tones). The brighter the red, the higher the density of languages with complex tonality, where the darker the blue, the higher the density of languages without complex tonality (Everett, Blasi, Roberts).

Morphological complexity is also dependent on the surroundings in which a language developed. Morphological complexity is how intricate the structure of words within a language is. It refers to elements such as affixation, or the addition of a group of letters to a root word to create a new word or new form of that root word, compounding, or combining two words to make a new word, inflection, or the changing of a word to reflect a distinction, such as tense, person, or number, and derivation, or deriving a new word from an existing word (eg. deriving impossible from possible). Languages with high morphological complexity tend to have many different forms of single words, whereas languages with lower morphological complexity tend to rely more on auxiliary words or the order in which the words are placed. For example, in Classical Latin, a language with high morphological complexity, single words can express more complex ideas than in English, a language with low morphological complexity. In English, we may say *we were working,* but in Classical Latin, one only needs to write *laborabat,* with little regard to word order in regards to a greater sentence. A direct negative correlation between morphological complexity and population has been observed in languages, which makes sense when regarding the pros and cons of languages with high or low morphological complexity. High morphological complexity can impact language processing, as more cognitive effort is required in order to learn and understand ideas in languages with high morphological complexity, making them much harder to learn. Contrast this with languages with low morphological complexity, which sacrifice wealth of information conveyed for more simple grammatical relationships within words, and therefore making a language easier to learn, and less difficult to create new languages from. It's no surprise, when considering this, that English has remained the lingua franca of the world, even considering England and America's colonial pasts and presents.

Thus, we can see how spoken languages have developed in response to their surroundings and the purposes which they need to fulfill.

Written language as compared to spoken language

Oral language developed organically and was shaped by the purposes by which it came by; it needed to be able to express ideas as efficiently as effectively as possible, taking into account population size, the environment itself, and the purposes for which its society used it. In contrast, the development of written language came about later in human history deliberately, in order to record information, preserve it, and transport it distances that oral language could not travel. Written language required the creation of symbols in order to represent corresponding sounds or words, which eventually led to the standardization of grammar, spelling, and punctuation. While oral languages evolved somewhat rapidly due to how often it was used and the need to keep adapting as humans themselves kept adapting, written languages evolved much more slowly due to grammar and other components being codified, the fact that written language does not require impulsive human interaction, and often the class restrictions to writing and reading.

One area where written language differs from spoken language is in its structure. While spoken language can rely on implicit things that are capable of being expressed in person, such as tone, body language, and rhythm, the written language requires much more explicit markers in order to convey the same meaning. This has resulted in the development of syntactic structures such as tone markers over certain symbols, punctuation, and even more modern conventions such as italics, bolded words, and underlined words. The need for this precision has also resulted in the development of more complex systems, such as the morphology previously discussed.

Abjad languages, or languages whose written forms do not include vowels, vs. languages that do include vowels, are examples of how written languages developed for certain purposes. For example, abjad languages' scripts are easier to learn, read, and write due to the omission of vowel sounds, which can often be quite complex. However, due to the ambiguity as a result of the lack of vowel sounds included in the written language, context and familiarity with the spoken language is necessary in order to truly understand the written language. As a result, non-Abjad languages are much easier in terms of language preservation.

Some written languages are deliberately preserved, such as the French language, which has an entire group of people dedicated to preserving "proper French", called the "Académie Française". This, therefore, results in a slower change in the French language, even in the spoken French language. However, due to the standardization of certain aspects of French, oftentimes, the way that colloquial French or certain French dialects are spoken is considered "improper" or "incorrect", despite simply being different ways of speaking the language itself.

Synthesis of Languages

The interactions between languages shape each language's structure, vocabulary, and phonetics. When different languages come into contact through various means, most often through colonization, proximity, or trade, they "borrow" different aspects of each other to fill "gaps" in them. This process of "borrowing" can lead to the diversification of vocabulary sometimes, however, it can also result in the loss of distinct vocabulary, especially when the culture the language belongs to has been colonized. For example, in Ireland, Gaeilge was the predominant language, however, once England forcibly imposed English onto the Irish, Gaeilge was replaced as the dominant language, and was changed to some extent. Additionally, languages can also adopt features from other languages due to contact. For example, the English language has incorporated numerous French words due to the Norman Conquest in 1066, resulting in English gaining up to 50% of its vocabulary, including words such as offspring, progeny, mansion, and pig. Due to sustained interaction between languages, noticeable changes in pronunciation patterns, grammar structures, idiomatic expressions and even writing systems can be observed, as well as new hybrid varieties of language, known as pidgins or creoles.

Pidgins and creoles are hybrids of languages that are a result of extensive interaction between language groups. Pidgins develop when people who speak different languages must communicate, such as for trade or labor, therefore resulting in a "hybrid language" consisting of simplified grammar and limited vocabulary made up of the most common/essential words from the pidgin's languages of origin in order to maximize efficiency and understandability. Over time, if a pidgin language becomes more and more frequently used, and is continued to be spoken as a primary way of communication, it can develop into a creole language with an expanded vocabulary and more complex grammatical structures. Creoles are spoken widely in a variety of different places across the world, and represent the lasting impact different cultures have on one another, whether good or bad.

Chapter Two
Impact of Language Loss and Preservation

When a language dies, so much more than words are lost. Language is the dwelling place of ideas that do not exist anywhere else. It is a prism through which to see the world.

Robin Wall Kimmerer

The Impact of the Loss of Language on Cultures

The extinction of a language is both an emotional tragedy and a cultural tragedy, as it signifies the loss of more than simply the words and grammar that constitute a language on paper. It also involves the disappearance of the unique worldview that a language encompasses, one that reflects the knowledge, experiences, and cultural identity of the people who used to speak that language. Language extinction affects specific cultures in ways that are difficult to explain, but that can destroy that culture's sense of self and identity, its history, and its place in the world. The loss of a language is the loss, to some extent, of a culture, and therefore the loss of human expression and thought in regards to how that culture viewed human expression and thought.

For many different cultures, language is inherently tied to identity. It carries the stories, the songs, the rituals, the collective memory, and so much more, of its people. It reflects its relationship with the world as a whole; the environment, its people's relationship with others, and the wisdom that has been gained throughout all of the years that culture has existed. When a language is lost, oftentimes it is akin to the soul of a community being lost. This is experienced on more than just a cultural level, but also on an individual level, especially those who may have been the last to fluently speak the language. Oftentimes, those last speakers experience incomprehensible grief as they witness the gradual erasure of their linguistic identity that they probably hoped to pass down to many future generations, but now are completely unable to, unfortunately.

For example, take the experience of many different communities. For those communities, language is central to their cultural heritage and their identity as people who belong in this world. In North America, many different indigenous languages have either become extinct, or are incredibly endangered. Each of these languages' express centuries if not thousands of years of history, and contain all of the unique forms of knowledge inherent to each tribe. For many of these people the loss of language can feel like the severing of the connection between the present people trying to keep their language and culture alive, and their ancestors who developed their languages and cultures. Language extinction, even more than that, can destroy oral traditions that have been passed down explicitly in their native languages, which are essential in order to preserve those cultures. In other words, the loss of a language is the loss of so many unique aspects of culture.

In many different indigenous cultures, the emotional impact of the death of a language is further intensified by the reason through which the language has died. Many such languages have been lost or endangered due to colonization, including forced assimilation policies, the suppression of indigenous expressions, and the mass death, whether intentional or not, of indigenous populations. For generations upon generations, indigenous children were forcibly removed from their families and sent to schools where they were punished for speaking their native languages, forced to adapt Western names, and often died due to harsh conditions or illness. The effects of this tragedy can still be seen today; some of these boarding schools, while they no longer have that purpose, still exist, indigenous children are still forcibly removed from their homes simply for practicing their cultural practices of living with many different family members or for unreasonable excuses, and indigenous children are still given Western names used outside of the home, and indigenous names used inside of the home and community. Another such impact is the fact that many communities now have very few fluent speakers left, and the younger generations often speak very little of their indigenous language, if they speak any of it at all. This forced alienation has affected these communities deeply, leading to feelings that many of us with immigrant roots can also relate to; the feelings of shame, loss, and displacement within their own cultures. The erosion of language under such circumstances is not natural; it was deliberate to erase indigenous identities in favor of more "civilized" western ones, and the results of that reflect that.

The emotional impact extends beyond the immediate community of speakers. Many indigenous people also feel a sense of mourning for their culture that is lost along with the loss of their language(s). For many members of the younger generation, there is a sense of guilt, or frustration, for being unable to speak the language that many generations before them could speak, even if the factors that resulted in that lack of being able to speak were outside of their control. Relearning these languages and repopularizing them is incredible, and deeply emotional, and involves not just relearning a language, but also reconnecting with a cultural heritage that was explicitly denied and destroyed for generations upon generations. Relearning these languages often results in cultural healing that is necessary for many people in order to live their lives fully and unashamedly.

The extinction of a language, more than that, affects how people interact with the environment to which the people of that language existed in. Many languages contain specific knowledge about the local ecosystems as a natural result of living there for thousands of years. As a result, those languages contain specific and necessary information that cannot be fully understood unless that language is understood. When a language is lost, that knowledge is lost too. A specific example of this knowledge's loss being extremely detrimental to the world as a whole is in regards to the California wildfires that have been occurring increasingly often over the years. The native populations in that area used to, and would recommend, setting smaller fires regularly to clear away the dead brush and to avoid larger fires on the scale that we have been seeing. Due to the endangerment of those cultures, the people now living there that aren't indigenous did not keep up with that practice, and now the wildfires in California are getting increasingly out of control.

Another example is the case of the Māori language of New Zealand. Although many efforts have been made in order to revitalize that language, it still remains on the verge of extinction, and its near extinction in the 1900s has had a profound effect on the Māori people. For many of the Māori people, the decline of their language reflected the marginalization and oppression they faced within New Zealand society. The loss of the Māori language was not just about the loss of communication, it was also about the weakening connection to the Māori people's spiritual beliefs, their historical narratives, and the lens through which they viewed the world. The revitalization efforts, which included the establishment of language immersion schools, have helped to increase how often Māori has been spoken, which has resulted in some much-needed reclaiming and healing for the younger generation of the Māori. The trauma of that near extinction still exists, however, and cannot be forgotten, and many elders still must live with the experience of having their language and culture forcibly taken away and almost killed for good.

Minority languages around the world are also at risk of going extinct due to the pressures of globalization and the dominance of major languages as lingua francas, including English, Spanish, Mandarin, and Hindi. In Europe, as an example, many different regional languages, such as Breton in France, Cornish in the UK, and Sorbian in Germany, are endangered as younger generations exclusively speak the national language for economic mobility, social norms, and because they no longer need to speak their dialects. The extinction of these smaller languages and dialects leads, often, to a sense of loss among speakers who have much of their identity tied to their language. Language is tied to more than just language; it's a tie to their folklore, culture, and customs, and the decline of many different smaller languages and dialects can feel like, and often is, the homogenization of different cultures in favor of larger and more dominant national or even global cultures.

The emotional toll of language extinction can also be felt in terms of the personal relationships that are lost with the decline of intergenerational transmissions. Language is, in many communities, a key component of familial bonds, which is used to more easily express intimate emotions from one generation to the next. More often than not, it is easier to express profound emotions in your native language, and as a result of the loss of your native language, often times a sense of isolation or emotional disconnect occurs as they realize their children and grandchildren cannot speak the language through which they express their emotions and their thoughts. This disconnect can create a sense of dislocation not only from the past but also from the present familial and communal life.

Despite the profound impact of the deaths of languages, many different communities fight this loss through efforts to revitalize language. While often incredibly challenging, these efforts can be a source of healing and community bonding. Revitalizing a language is about more than just preserving the syntactic structure of a language; it's also about restoring community and cultural pride, reconnecting with your roots, and understanding the generational knowledge that has been within families and cultures for thousands of years. The process of revitalizing languages can be deeply emotional, and thus it is imperative that we go to great lengths in order to streamline the process as much as possible, make it accessible to as many people as possible, and make it as effective as possible. It isn't enough to just teach younger generations how to speak a certain language; we must find ways in order to preserve that language as it is in case it is impossible for it to be fully recovered; that way, the knowledge of that culture is never truly lost, and we can honor the generations that have come before us.

The Impact of the Loss of Language on Global Knowledge

The extinction of a language is not merely a cultural or emotional event but a profound loss to the entire body of human knowledge. Every language encapsulates a unique way of understanding the world, from ecological insights to social structures, medical practices, and philosophical perspectives. The death of a language is, in essence, the disappearance of an entire system of knowledge, much of which cannot be replicated or fully understood through other languages. The global impact of language extinction extends far beyond the affected community, as it represents the erosion of human diversity in its most intellectual and practical forms. The consequences touch on everything from scientific research to environmental conservation, to how societies understand and approach problems.

One of the most significant losses associated with language extinction is the disappearance of traditional ecological knowledge. Many indigenous and minority languages contain detailed, place-based understandings of local ecosystems, plants, animals, and natural cycles. This knowledge, accumulated over centuries of interaction with specific environments, is often embedded directly in the language itself. For instance, certain indigenous languages contain vocabulary for plants or animals that do not exist in any other language, along with information about their uses, growth cycles, and the ecological relationships between species. When these languages die, the knowledge they carry about biodiversity, sustainable practices, and the environment may be lost as well.

Numerous languages that are spoken by the indigenous peoples in the Amazon rainforest region contain vast amounts of knowledge about both the plants in that region and their usage. Many of these plants have not been studied by Western science, as their knowledge has been ingrained in the indigenous cultures and it was not necessary to write down their purposes. In the absence of the language that encodes this knowledge, much of it disappears. This is not just a loss for the local community but for global medical research, as many of these plants could hold the potential for new drugs or treatments for diseases. Without the specific linguistic and cultural context that comes with the language, outsiders may not even know where to look or how to properly identify and utilize these resources.

Languages also encode unique taxonomies for categorizing and understanding the natural world, which can differ significantly from the scientific classifications commonly used today. The taxonomies of indigenous people are more nuanced, recognizing distinctions between different factors of the ecosystem in which they live that go unnoticed by those from outside those communities, due to the fact that those indigenous people have been living in those areas for unfathomably many years. For example, the Yupik people of

Alaska, like the Inuit people, have a vast number of words for different types of snow and ice, due to the fact that they spend much of the time surrounded by snow and ice. Each term provides information about that snow and ice, including the texture, density, and stability of it, demonstrating the intimate knowledge of that environment. The loss of that vocabulary would result in the loss of incredibly specialized knowledge in how to survive in such an unforgiving climate, and loss in efficiency in expressing how to survive in specific high-stress situations.

Another area in which the loss of language impacts the loss of knowledge globally is in regards to the fields of anthropology and in regards to sociology. Languages serve as a lens through which societies structure themselves and make sense of the world, and the grammar, syntax, and vocabulary of a language reflects that society's values, priorities, and social systems. For example, languages from smaller more egalitarian societies often do not use hierarchical forms of address, which reflect the fact that egalitarian societies prioritize equality and do not have as solid of a hierarchy. Other languages may have complex systems of respect embedded within their grammar, offering insights into how social hierarchies and relationships of power are constructed and maintained.

The loss of these languages means the loss of these social insights. Those who study social insights rely on language as a tool for understanding how different cultures organize social systems, kinship structures, and their modes of governance. If a language is lost, the knowledge that anthropologists and sociologists need becomes inaccessible or is not thorough enough to reflect the actual complexity of the societies to which they belong, a tragedy in and of itself. The extinction of a language thus represents a narrowing of our understanding of the diversity of human societies, their values, and their ways of living.

The death of a language also hampers the efforts in the study of language in order to understand the full range of human cognitive capabilities. Each language can present a different model on how the human brain functions. Therefore, by studying the world's different languages, linguists have been able to uncover and understand how humans have adapted language in order to suit their needs in a wide variety of environments. Therefore, the extinction of languages diminishes the variety of data available to linguists which, in turn, limits our collective understanding of how human beings function and have functioned throughout the history of humanity.

In addition to the structural aspects of language, the stories, oral histories, and traditions embedded within languages also represent a vast repository of human knowledge. Many languages that are now endangered or extinct were primarily oral, meaning that the stories and histories they carried were never written down. These oral traditions often contain important historical information, including migration patterns, interactions with other cultures, and events that shaped the course of human history in specific regions. For instance, Polynesian navigational chants encode detailed information about ocean currents, wind patterns, and stars that were used for long-distance sea travel. Without the language, much of this knowledge has faded, making it difficult to fully understand how ancient Polynesians managed to explore and settle islands across the vast Pacific Ocean.

The death of a language would also mean the loss of the ethical and metaphysical knowledge that differs from culture to cultures. Every language can provide a lens through which to view complex questions in regards to life and death and the afterlife. The linguistic way by which a language speaks about philosophical concerns reflects the people who speak that language. For example, some languages have specific words or phrases for types of relationships/ethical dilemmas that don't have exact translations in other languages. The disappearance of a language means the disappearance of these frameworks for understanding and approaching philosophical questions. This diminishes the diversity of human thought and reduces the range of perspectives that are available when grappling with ethical challenges in a globalized world.

In addition, language extinction has serious consequences for historical knowledge. Many languages, particularly those of minority and indigenous groups, carry within them records of historical events, social changes, and interactions with other cultures that may not exist in written form elsewhere. When a language dies, these historical records are often lost with it. This is problematic especially in regions where most of its history is passed down orally, or when the dominant culture has overwritten the history of marginalized groups. For example, much of Norse and Irish culture has been "christainified" due to the influence of the dominant Christian culture, therefore resulting in much of the mythology having retroactively added Christian terminology that did not accurately reflect the culture from the time. The loss of those languages results in the loss of differing historical perspectives, resulting in an incredibly biased viewpoint on history that often does not accurately represent the full scope of the historical event that is being portrayed.

The death of languages has broad implications in terms of global diversity and global innovation. Human progress has been fueled by the expression of ideas and the ability for us to approach problems from unique angles dependent on our backgrounds. Thus, the diversity of languages and cultures has often been a key factor in successful innovation. Thus, when languages die, and people become more homogenous, innovation slows down. This can have detrimental effects in fields where innovation is incredibly necessary, such as STEM realms, and must be rectified.

Moreover, language extinction contributes to the homogenization of global culture. As major world languages like English, Spanish, and Mandarin dominate, smaller languages are increasingly marginalized. The result is a narrowing of cultural expression and creativity. The stories, songs, and traditions that are inherently tied to language also end up being lost, thus reducing the richness and diversity of the global culture in which they reside. This loss of cultural diversity is both an inherent tragedy and also demonstrates a weakening in the resilience of human societies. The more homogenous the world becomes, the more we are unable to adequately respond to unique problems that can only be solved through the collaborations of different cultures.

In conclusion, the extinction of languages represents, and results, in an incredible loss in the world's collective knowledge. Each language is unique, and therefore represents a unique lens through which to view the world, and therefore the loss of even one singular language represents the loss of knowledge on a global scale. More than one language dies out each week. This is a tragedy that must be prevented through the process of language preservation.

Physics and the Loss of Language

Language is a dynamic and evolving system, shaped by many different cultural, social, and technological forces. Over time, all languages undergo changes that result in the loss of certain words— something known as lexical attrition. These words are often either replaced, or even lost entirely, due to shifts in the society that the language belongs to, whether that be due to the evolving culture, technological advancements that result in less of a need for a specific word or result in the changing meaning of a word, or simply due to the linguistic structure of the language itself changing. The disappearance of words can have fascinating, and oftentimes concerning, implications for both intra- and inter-cultural communication, especially when considering the complex relationship between language, cognition, and the perception of reality.

One of the main ideas when considering the impact of the loss of words on communication is called the Sapir-Whorf hypothesis. This posits that the language that we speak can influence our view of reality. When a word ceases to be used, it oftentimes demonstrates a resulting shift in how speakers of a language view or think about certain aspects of the world itself. When words related to specific cultural practices are lost, for example, the underlying concepts that are represented by these words may also fade from that society. This is particularly problematic in the context of scientific vocabularies, where the loss of precise terms can lead to confusion within that field and the loss of the way the concept represented by that term is taught.

The precise language used to describe phenomena is crucial in order to convey nuanced ideas and to maintain the integrity of scientific discourse. As language continues to evolve more and more, and as words are lost, the scientific community can face challenges in preserving the clarity and precision necessary for efficient and effective communication. For example, scientific texts that are older contain terminology that is no longer in use, or that have been replaced by more modern, accurate terms. Thus, the loss of these words can make it more difficult for contemporary readers to grasp the ideas expressed if no context is provided, thus resulting in the need for reinterpretations.

An example of a term falling widely out of use is the term "aether", which was once widely used in order to describe the medium through which light travels. The concept of "aether" was central to many physical discussions up until the early 20th century, when experiments, like the Michelson-Morley experiment, demonstrated that light did not require a medium for propagation, thus leading to the abandonment of aether. As the concept was discarded, the term fell out of use, marking a significant shift in how physicists conceptualized space. The loss of the term "aether" was a natural result of developing sciences, however, it does represent how terms disappearing mark significant changes in scientific discovery. The discarding of the term "aether" results in the need to explain the meaning of that term to younger generations, as well, something that is not necessarily negative, but that also does express how the realms of science change with the times.

The impact of lexical attrition extends beyond the loss of specific technical terms. It can also affect how we communicate about abstract concepts. For example, in Old English, the word "wyrd" refers to a concept akin to fate, and was deeply intertwined with cultural and spiritual beliefs of the time. As English has evolved, the word "wyrd" fell out of use in English, and the concept, either as a result or as a cause, represents the decline of that word. The loss of that word illustrates how the disappearance of a word can result in a lessened capacity to express certain ideas, particularly in regards to those that are essential to a specific cultural worldview and express ideas central to that specific cultural worldview.

In the context of physics, the loss of words related to philosophical or metaphysical concepts can similarly impact communication and the development of new theories. Physics, particularly in its more theoretical domains, often deals with abstract concepts that require precise language for effective discussion. When words that capture specific nuances or dimensions of these concepts are lost, it can lead to a reduction in the richness of the discourse. This, in turn, may constrain the ability of physicists to explore new ideas or to build upon the work of their predecessors.

An incredibly significant implication in regards to the loss of words in languages is the potential of narrowing our frameworks in terms of cognition and perception. Language is capable of shaping how we categorize the world as a whole, and the loss of certain words can result in a corresponding loss in the diversity of our thought. In information theory, this is seen as a reduction in the linguistic entropy of language, a concept that will be discussed later in this book. As languages lose words, the range of possible thoughts and ideas that can be expressed within those languages may also diminish.

Moreover, the loss of words can affect other aspects of language, such as syntax, semantics, phonetics, and more. As certain words stop being used, the structures that depend on the usage of those words can also change, thus resulting in broader shifts in terms of the language as a whole. Thus, the precision and unambiguity of communication can be affected, especially in regards to fields that rely on highly specialized vocabularies, including physics. For example, the loss of some adjectives or adverbs that at one point allowed for specific distinctions may result in a more uniform, homogenous, language, which makes it even harder to convey certain subtle differences.

The phenomenon of lexical attrition is also a contemporary concern, as languages around the world continue to face increasing pressures of globalization and the increased usage of certain technologies. The dominance of English as a global language, combined with the rapid pace of the advancement of technology, has resulted in the decline of many different regional and minority languages. As these languages continue to disappear, unique words and concepts that they contain also disappear, therefore resulting in the loss of those cultures. These losses are tragedies, and should be treated as such.

The challenge of preserving linguistic diversity is closely linked to the need to have precise communication. As discoveries are made and concepts are developed, there is an increasing need for language to evolve to the changing times, and perhaps even lose some of its non-useful terminology. This evolution, however, must be balanced with the preservation of its linguistic resources, especially those that are important culturally. The loss of words in scientific language can lead to a diminished capacity for interdisciplinary communication, as well as a potential loss of cultural heritage.

The loss of words over time is a natural, and expected, part of languages as they grow and develop, but still, it carries certain implications for communication, especially in regards to fields that require precision in their terminology. The disappearance of some words can also lead to a narrowing of cognitive frameworks, therefore homogenizing thought and even further limiting the ability to express unique, complex, and emotional ideas. These phenomena underscore the importance of preserving linguistic diversity, and also maintaining a balance between the evolution of language and the preservation of its richness. As we continue to explore the intersections between all of these seemingly disparate fields, we must remain mindful of the real impact that language has on the world itself and our understanding of that world, and we must also consider the impact of lost words on our ability to communicate effectively.

Linguistics and Physics Development and Processes

Audio Synthesis

Audio synthesis is the process of creating sound through electronic means, using digital or analog devices. This field encompasses a wide range of techniques as well as many different technologies, including basic waveforms and oscillators as well as sophisticated signal processing algorithms, such as the fast Fourier transform (FFT) filter, which selects frequency components from a signal, and the finite impulse response (FIR) filter. The physical processes used in audio synthesis involve the manipulation of electrical signals, acoustic properties, and psychoacoustic phenomena to produce a rich variety of sounds, which can then be used in many linguistic processes as well as in language preservation.

Audio synthesis is built around the generation of waveforms, or graphical representations of a periodic signal. These can take on many distinct forms, which include sine waves, square waves, sawtooth waves, and more. Each unique type of waveform has distinct characteristics that can therefore be utilized to create different timbres or tones.

A fundamental physical process in audio synthesis involves generating waveforms using oscillators. An oscillator is an electronic circuit that produces an alternating current (AC) at a specific frequency determined by its control voltage input. In the context of audio synthesis, oscillators create periodic waveforms at audible frequencies (usually between 20Hz and 20kHz). These oscillating signals form the basis for creating musical tones across various instruments and genres, as well as other sounds.

In analog synthesizers, oscillators often rely on voltage-controlled circuits to produce continuous variations in frequency based on input control voltages. By modulating these control voltages using various sources such as envelopes or low-frequency oscillators (LFOs), synthesists can create dynamic changes in pitch over time—enabling expressive musical articulations like vibrato or pitch bends.

In digital synthesizers and software-based synthesis systems (such as soft synths), oscillator functions are typically implemented with mathematical algorithms—referred to as wavetable synthesis—which calculate discrete samples representing the desired waveform at precise intervals based on user-defined parameters.

Once generated by an oscillator, these raw waveforms undergo further manipulation through processes like filtering—a critical aspect within audio signal processing employed for shaping tonal qualities by attenuating certain frequencies while allowing others - thus altering harmonic content. Low-pass filters are commonly used in subtractive synths for removing higher harmonics; emphasizing fundamental frequencies- yielding warmer timbral character; whereas high-pass filters diminish lower frequencies enhancing brighter tonal hues.

Moreover, amplitude modulation – another crucial method - allows controlling output volume via manipulation level modulated signal affecting overall loudness. Envelopes often used shaping amplitude profiles- encompassing phases like attack - influence speed onset sound, decay – rate reducing volume, sustain maintaining level & release determining duration tone lingers after key release.

Beyond traditional subtractive methods other approaches involving additive/subtractive/frequency modulation/ phase distortion offer diverse timbral possibilities – enabling synthetic emulation natural acoustic sonority across instrumental spectrum; broadening creative horizons facilitating novel sonic exploration benefiting studio productions live performance scenarios.

Furthermore, spatial effects offered manipulating timing and intensity reflections emitted soundwaves interacting environments e.g reverb, and delay altering perception distance proximity within the auditory field enriching immersive listening experiences for musicians and audiences alike. These technologies significantly impact compositions enhancing emotional resonance delivering captivating performances transcending conventional artistic realms, a topic that was discussed more thoroughly earlier in this book.

Psychoacoustic principles also play a crucial role during the development of audio synthesis applications in exploring human auditory perception discerning parameters governing phenomena. Critically judging synthesized sounds for authenticity acceptance among listeners validates effectiveness of innovations and breakthroughs by industry influencers keenly monitoring creative progressions competitive edge harnessing cutting-edge advancements exceeding audience expectations standards fostering unwavering relevance dynamic industry landscape ensuring sustainable growth continuity innovation evolution.

Information Theory

Information theory is an incredible framework often utilized for understanding things related to information; particularly its transmission, storage, and processing. One of information theory's core tenets is to provide a mathematical foundation for quantifying the amount of information in a message. Coupled with this is actually understanding how that information can be processed and then transmitted as effectively as possible. It may seem like a concept purely pertaining to computer science, engineering, and other math-heavy STEM fields, however, its usage in physics as well as linguistics is an often overlooked, but incredibly important.

In physics, information is tied to the fundamental laws that dictate the universe. Physicists have long grappled with questions pertaining to how information is encoded in physical systems, including on a macro and subatomic level, and how that information can be manipulated.

One large connection between information theory and physics itself is through entropy. Entropy in the physical sense is the measure of randomness in a system. For example, no chemical reaction can ever occur that decreases entropy in the universe, as each chemical reaction contributes to the increasing randomness of the universe. Entropy is also related to the amount of uncertainty in a system's subatomic state, as we exist on the macroscopic level and any observations that can be made must alter that system's state on some level, as the mere act of observation results in a change in that system. Thus, entropy reflects a core tenet of information theory; uncertainty can be defined as the missing aspects of information.

Physicist-turned engineer Claude Shannon's work on information theory laid out this connection by defining entropy as a measure of the average "surprise" in a system, thus capturing the uncertainty pertaining to what we will observe next, or, in another light, what we need to know in the future. This insight thus allowed Shannon to utilize mathematical principles from probability that relate directly back to concepts such as thermodynamic entropy, thus streamlining that communication rather efficiently.

By viewing information theory through the lens of physics as a whole, we can also view it as extending beyond purely informational domains and rather into intricate physical systems. This thus opens up possibilities in other fields, such as in astrophysics, quantum mechanics and computing, and other less intuitive fields such as linguistics, as is discussed in this book. All of these pursuits investigate the underlying nature of reality itself, thus encouraging breakthroughs on an unprecedented level.

When viewed through the lens of quantum mechanics, information theory offers a rich tapestry that explores the intricate connections that entangle particles across vast distances, a concept cleverly named entanglement. This enables the emergence of breakthrough paradigms thus encouraging innovative technologies in regards to classical computation. Viewed through this lens, information theory serves as something that can be a major breakthrough in regards to both the fields of information theory and quantum mechanics

When viewed through the lens of classical mechanics, information theory provides insights regarding topics such as the conservation of energy, matter, and other topics that govern the physical processes that shape the universe itself, thus revolutionizing the field of classical mechanics. Furthermore, when considering singularities such as black holes, a situation where information paradoxes arise due to the conflicting nature of quantum mechanics, general relativity, and classical mechanics, information theory provided a pivotal role in preserving virtual records accumulated throughout the universe's lifespan. This lends itself to some more philosophical questions, such as whether or not our lives as human beings will be preserved, and how.

Physics and Language

Understanding why certain sounds/words are used regularly even across different languages requires delving into the physics of sound, particularly its production and perception. The way that some words are formed, especially those pertaining to universal human experiences like familial relations, acquiring food, and human connection can be explained somewhat through the study of the physical properties of sound, such as certain sound's frequencies (Hz), their resonance, and the way humans actually produce sounds. In order to analyze this connection, we must use concepts like vocal frequencies, spectrograms, and the acoustic properties of sound in order to understand why certain phonetic patterns have emerged across many distinct languages.

Human speech is produced by various parts of apparati interacting, therefore allowing us to speak. These apparati include the lungs, vocal cords, mouth, and nasal passages. When speaking, air is pushed through the lungs into the trachea, resulting in the vibrations of our vocal cords. As a result, sound waves are generated. The distinct sounds of our voices are the result of certain characteristics, such as the tension in our vocal cords, their length, the shape of our vocal tracts, and the air passages above our vocal cords.

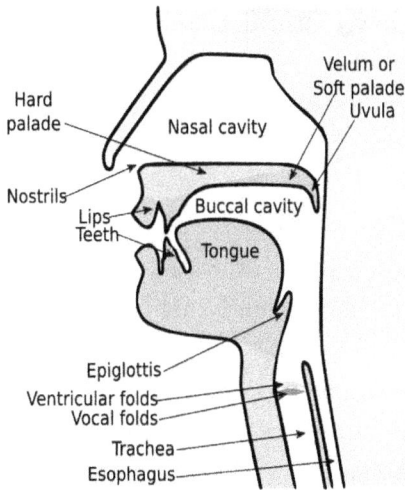

A schematic diagram of human vocal apparati

The fundamental frequency, with an SI unit of Hertz (Hz), is the lowest frequency of a resonant frequency, which is the natural frequency vibration, which can be determined by the physical features of the vibrating object. In this case, the vibrating object is the vocal cords of the human who is speaking. Adult biologically male voices have a typical fundamental frequency ranging from 85 to 180 Hz, while adult female voices usually range from 165 to 255 Hz. Infants and children tend to have even higher frequencies due to possessing vocal cords that are even shorter and thinner.

Speech sounds are also characterized by formants, which are resonance frequencies of the vocal tract. Formants are essential to determining the quality of vowels and some consonants. The two formants, F1 and F2, with F1 formants being sounds with frequencies under 1,000 Hz and F2 formants being sounds between 1,000-2,000 Hz, are important for vowel identification in particular. For example:

- The soft a sound as in "father" usually has a low F1 (about 500 Hz) and a low F2 (about 1000 Hz).

- The hard e sound as in "see" has a low F1 (about 300 Hz) and a high F2 (about 2500 Hz).

Consider the acoustic properties of some words that make use of universal sounds, such as those for "mother." The usage of bilabial consonants ([m], [p], [b]) and open vowels ([a]) in words for mother across languages can be explained by their acoustic properties and ease of production.

Now, consider the acoustic analyses of some of the sounds most common in the words used for mother

[m] Sound:

• The [m] sound has a low frequency with most of its energy concentrated in the lower frequencies, around 100-300 Hz. This low-frequency sound is often perceived as soft and comforting, traits that are often associated with being a mother.

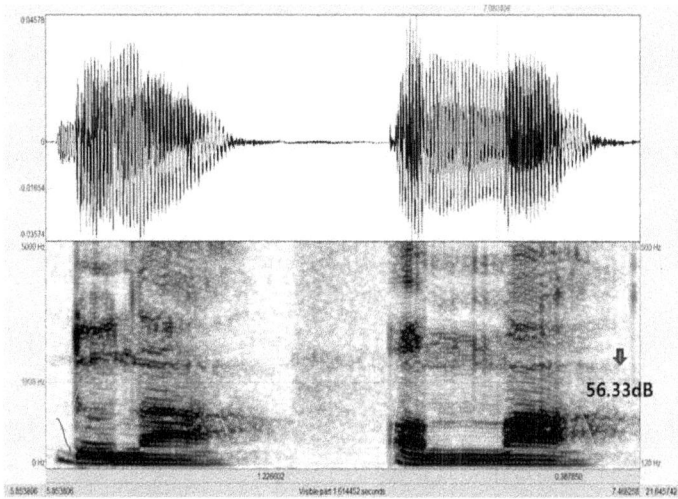

A spectrogram of the m sound present in the word mother shows a lot of energy in the lower frequencies, with a gradual drop-off as the frequency increases.

"ah" Sound:

- The vowel a, as is present in the words "mama" and "amma" has a characteristic F1 formant around 500 Hz and an F2 formant of about 1000 Hz. This makes it an "open" vowel acoustically, which is not only easy for infants to produce, but is also easy for adults to recognize.

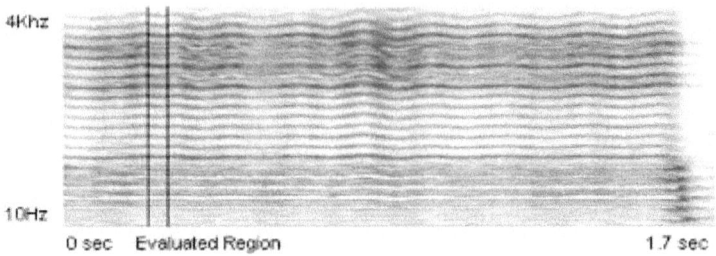

4Khz

10Hz

0 sec   Evaluated Region   1.7 sec

Spectrogram: The spectrogram of the "a" sound in "mama" shows prominent energy in both F1 and F2, reflecting its open quality.

A spectrogram is a visual representation of the "loudness" of a signal over time across various frequencies. By using a spectrogram, you can see if there is more energy at one frequency rather than another, and how that energy varies over time. In the context of language formation, spectrograms can also be used to compare the acoustic signatures of different sounds across languages. In terms of language preservation, spectrograms are also a way of clinically preserving language for future generations.

By analyzing spectrograms of the word "mother" in different languages, we can observe that despite differences in certain phonetic details, the acoustic pattern remains somewhat similar. This consistency supports the idea that certain sounds are universally preferred for expressing some fundamental concepts, such as ones relating to familial bonds.

- English "Mother": The spectrogram shows a strong bilabial closure at the beginning for [m], followed by the open vowel [ʌ] with clear formants.

- Tamil "Amma": The spectrogram would be similar, showing the bilabial [m] twice with an open [a] vowel in between, producing a rhythmic pattern that is easy to produce and recognize.

The human vocal tract acts as a resonant cavity, or a cavity that is capable of producing a certain resonance pattern. This is why vowels, which are primarily capable of being differentiated by their formants, have distinct acoustic properties that are easily recognizable across different languages, even if the vowel sound in one language sounds similar to the vowel sound in another.

- Open Vowels: Vowels like [a] resonate well within the vocal tract, producing strong formant frequencies that are easier to distinguish than some other ones. This makes them ideal for basic, universally understood words.

- Closed Vowels: Vowels like [i] or [u] have higher formants, which make them less universally favored for basic words.

Infants naturally produce certain sounds, such as [m] and [a], due to their simpler motor requirements. These sounds are then reinforced by those that surround the infants, leading to their usage in the words for "mother" and other universal concepts.

Language is a product of both physical constraints (such as the acoustics of the vocal tract, the length and thickness of vocal cords, and the ways that other parts of our vocal apparati are shaped) and cognitive constraints (such as the brain's ability to process and interpret sound which can vary with age and overall cognitive function). The universality of some sounds in certain basic words reflects the intersection of these constraints, thus resulting in the usage of similar words across multiple different languages.

The physics of sound offers a perspective on why some phonetic patterns have become prominent across different human languages. The intersection between physics and linguistics is something that is not well explored, yet has many applications especially in regards to the preservation of language. Many tools, like spectrograms, oscilloscopes (tools that show the voltage of a signal over time), and goniometers (a tool that can show the relationship between the phase of multiple signals) are, and will continue to be, necessary in order to further study language preservation and the intersection between physics and linguistics.

The evolution of language is a process that is influenced by many different factors, including in regards to social dynamics, cognitive mechanisms, and environmental changes. While traditionally studied within the realms of both linguistics and also anthropology, physics has quite a bit of potential in regards to being used to predict language evolution. This oft understudied realm is an emerging area of interdisciplinary research that will, perhaps, revolutionize both fields. This approach will draw on concepts involving statistical physics, information theory, as was previously discussed, and dynamical systems, thus offering fascinating insights into how languages may evolve as time passes, and hopefully revolutionizing both fields to great extents.

A main idea borrowed from physics to study language evolution is dynamical systems. In physics, these systems are mathematical models that describe the state of a system over time, represented by differential equations most of the time. Language can be viewed as a dynamical system in which the rules that govern its mathematical system are, instead, rules that govern language itself, such as grammar, convention, syntactic structures, and more, that lead to changes in the language over the time. These changes can also be influenced by internal factors that therefore must be considered, such as the bias of speakers and listeners, the external environment in which the language is located, and social and cultural interactions.

In order to predict how language can evolve, researchers can model the change of languages by using mathematical models of motion. For example, a model could treat the frequency of a linguistic element, such as a word or syntactic convention, as a variable that could change over time. The rate of change of this variable could be influenced by factors that must be taken into account, such as the commonality of the usage of the variable itself, the ease of its usage, and the efficiency of its communication. These influences must therefore be modeled as forces acting on that linguistic variable, similarly to how forces in physics result in changes in the motions of objects.

Returning to statistical mechanics, the idea of phase transitions can be applied to language evolution. In physical systems, phase transitions occur when a system undergoes a sudden change, such as when a liquid reaches its critical temperature and transitions from water to steam, or when a protein reaches its critical point and begins to unfold. A phase transition, when applied to language, could represent a sudden change in a linguistic variable, such as the adaptation of a new word or the loss of an old one. Some examples of this are the shift from the usage of "thou" to "you", or when English became "latinified" and changed words like "ile" to "isle" in an attempt to "Romanify" English. A more modern example of a linguistic phase shift pertains to trends; words that were not used often in the past few years skyrocket to usage, such as the words "slay" or "demure", as they become trending, and then are just as quickly dropped. No one really uses the slang word "yeet" anymore, despite its widespread usage amongst the youth only a few years ago.

Agent-based models are another concept based from physics that can be used to simulate language evolution. Agent-based models simulate a large number of individual "agents", each of which represent a common variable. The agents interact with each other based on specified rules. In linguistics, these agents could be a speaker of a certain language, and they could interact according to specified rules, such as the likelihood of adopting a new linguistic phrase based on several different factors, such as its usage online, in communities, and the ease of its use. Over time, the aggregate behavior of the agents within this system can reveal patterns of linguistic change, such as the spread of those new words or exceptions to linguistic rules. This approach could be seen as analogous to modeling the behavior of systems, such as the particles in a gas or liquid, where interactions between particles can result in macroscopic properties like the temperature or pressure in a certain area.

Information theory can also offer an interesting framework through which we can understand the evolution of language. Information theory, which was originally discussed by Claude Shannon, someone who was previously mentioned earlier in this book, developed it in order to discuss communication systems and information. Information theory provides tools to quantify the amount of information contained in a linguistic system and how efficiently it could be transmitted.

Over time, languages need to evolve in ways that balance efficient communication and ambiguity. This balance is achieved through the usage of information-theoretic principles. For example, as language becomes more regularized, its entropy may decrease, thus reflecting a decrease in exceptions to rules and the combination of distinct forms in order to reduce the amount of those forms. If language, however, becomes too regular, it may lose its ability to express subtle distinctions that result in less efficiency, but more clarification and high range of expression. Thus, information theory can be utilized to model this balance between both efficiency and clarity.

An incredibly fascinating aspect of the usage of physics in order to predict language evolution is the potential to identify the absolute laws that govern linguistic change, some of which were discussed earlier in this book. Physics is governed by universal laws, such as the theorem that matter and energy cannot be created or destroyed, and that an object in motion must stay in motion unless an outside force is applied, linguistic systems might be governed by universal principles that can dictate how languages are able to change over time. An example that was discussed earlier was Zipf's law. Understanding the underlying themes that gave rise to Zipf's law could theoretically provide insights into the general principles that govern language evolution.

Network theory from physics is also applicable in the study of language evolution. For example, languages can be represented as networks with nodes corresponding to linguistic elements, like words or sounds, and edges can represent the relationships between them, such as being homophones or similarities in the way they are spoken. These networks can also be analyzed in order to identify key nodes or clusters that can play a central role in the language that is being depicted in this theory. Over time, changes that appear in the network can reveal patterns of how that language itself changes.

Simulations that can combine these physical principles can provide a powerful way of predicting how language evolves. By creating models that combine aspects of dynamical systems, statistical mechanics, information theory, and network theory, researchers can as a result simulate the long-term evolution of language in response to different situations. These simulations are able to test hypotheses about factors that drive linguistic change, such as social influence, technological innovation, or the population of the culture in which the language is used. The results of these simulations can then be compared with past data in regards to that language in order to both validate the simulation as well as to refine the predictions made because of the simulation.

In conclusion, physics can be applied to the study of language evolution, resulting in a promising interdisciplinary approach that can utilize both the mathematical rigor and the predictive power of physical equations in order to understand the complex dynamics of changing languages and as a result shed new light on both physics and linguistics.

Linguistic Theories and Quantum Mechanics; an Intro to Using Physics for Language Preservation

Quantum physics and linguistics, which initially appear to be two completely distinct fields with very little overlap, have somewhat similar fundamental processes. Quantum mechanics contains principles like superposition, or the ability of subatomic particles to exist in multiple states until the moment of observation, entanglement, or how some particles remain connected, or "entangles", no matter their distance, and uncertainty, or the inability for all aspects of subatomic particles to be observed at once since the mere act of observation results in a change in the state of the particles. These principles provide a framework that has inspired a few approaches in linguistics, as well as in cognitive science. These ideas are new and mainly speculative, but they explore how some prominent quantum physics principles might, either metaphorically or even physically, explain the structure of language and cognition in new and innovative ways.

A main area that has overlap between quantum physics and linguistics is information; something that linguistics studies the conveying of, and something that quantum physics has streamlined in many different ways. In quantum physics, for example, information is treated as a fundamental aspect in most physical systems, which is demonstrated in the state of many subatomic particles. In linguistics, for example, language serves as a system used in order to encode, transmit, decode, and analyze different pieces of information. These parallels between linguistics and quantum physics suggest some sort of overlap between these two fields, and can be used to revolutionize them. A prominent idea that synthesizes the intersection between these two fields compares the ambiguity of language to the ambiguity of language. This idea is that a quantum state, which represents many multiple possibilities simultaneously occurring, is comparable to the ambiguity of language, in which a single sentence, or word, can convey many different meanings until the context in which the word or sentence is placed clarifies the true meaning of it. Thus, the ambiguity of language can be considered to be analogous to the collapse of a quantum wavefunction.

This is an introductory idea in the principle of quantum linguistics, where linguistic researchers use quantum mechanics, which is quantum physics formulated into mathematical laws, to model many of the complexities of human language. A prominent example of this is the linguistic application of the physical superposition principle, which has been employed to represent the coexistence of multiple meanings in a word or phrase. By using the idea that something can be in multiple states until the time it is observed, we are able to address complex linguistics ideas such as context dependence, and the ambiguity of a word until its actual use in a clearer way than is possible using traditional models of language. Another prominent example is the principle of entanglement, a phenomenon where particles become correlated in such a way that the state of one instantaneously influences the state of another. This has been used to describe the interconnectedness of words within a sentence or a broader linguistic context. For example, in the phrase "bated breath", bating has a meaning, whereas it does not exist in conventional speech outside of the word "breath", making its state dependent on whether or not "breath" is present.

Another area where physics and linguistics have an intersection is in cognitive science, where concepts such as someone's perception of something can demonstrate the usage of both physics and linguistics. Some research suggests that quantum physics plays a role in cognition, particularly in regards to decision-making or pattern recognition, two things that are fundamental to language processing. The idea here is that the brain could use quantum computation in order to solve certain complex problems more effectively than classical computation could allow. If this is true, it implies that linguistic processing may also theoretically be influenced to some extent by quantum effects, though this remains both highly speculative and controversial. More research is required before a true cause and effect can be established.

The idea of quantum cognition, or the usage of quantum ideas in order to explain aspects of psychology, such as human judgment and decision making when traditional methods fail, also demonstrate a further connection between quantum physics and linguistics. For example, the probability amplitudes in quantum mechanics, which determine the likelihood of different outcomes, can be used to model how certain humans weigh different interpretations in language. This approach offers a way to model the non-linear, context-dependent nature of language and thought in a way that many approaches do not consider.

Despite these intriguing possibilities, it is important to note that the convergence of quantum physics and linguistics remains largely theoretical and metaphorical at this stage. There is currently little to no significant empirical evidence to suggest that quantum mechanics directly governs linguistic processes. However, the application of quantum physics in modeling some complex linguistic phenomena shows the potential for cross-disciplinary innovations. The mathematical tools of quantum mechanics are capable of providing novel insights into the structure of language, similarly to how quantum mechanics has revolutionized our understanding of the physical world.

The philosophical implications of quantum physics, particularly the ways in which it challenges classical notions of reality, determinism, and locality, and have resulted in many theories in order to reconcile quantum physics with those classical notions, resonate with certain linguistic theories that emphasize the fluid and dynamic nature of meaning and the interpretation of different languages and linguistic phenomena. The indeterminacy and context-dependence of quantum states echo the flexibility and context-sensitivity of language, where meaning is not fixed, but rather emerges from the interaction of words, speakers, and listeners and the context in which it ultimately lies.

In conclusion, while the direct convergence of quantum physics and linguistics is still theoretical at most, the interdisciplinary dialogue between these fields holds promise for advancing our understanding of both language, cognition, and more. The usage of quantum concepts in linguistics, particularly in the realms of information theory and cognitive science, offers a new lens through which to explore the complexities of human communication. As research in quantum cognition and quantum linguistics progresses, it may yield even more useful models that serve to challenge our traditional views and may open even more new avenues for exploring the nature of language.

The intersection between quantum physics and language preservation, which also seems like a unique intersection with very little overlap, offers fascinating potential for addressing an incredibly salient issue in linguistics: the loss of linguistic diversity. Every day, multiple languages face extinction, and the application of quantum mechanics may provide unique solutions for preserving languages as a whole. By leveraging quantum concepts like quantum computing, quantum cryptography, and the principles of quantum information theory, researchers may develop even more new methods in order to utilize linguistic data with increasing access, security, and efficiency.

Quantum computing, a field that utilizes the principles of quantum mechanics, has the potential to revolutionize the way we understand linguistic data. For example, classical computers process information binarily, which means that it processes data that has been broken down into 0s and 1s. On the other hand, quantum computer utilize something called qubits, which utilize the concept of superposition in order to have the capacity to represent both 0s and 1s simultaneously. This allows quantum computers to perform complex computations much faster than binary computers, thus making them well suited to tasks that use large amounts of data, such as the documentation of an entire language, thus allowing those languages to be preserved.

One challenge in language preservation is the vast volume of linguistic data that must be recorded, analyzed, and stored. Many endangered languages, most of which are spoken by smaller communities, use oral traditions, or traditions passed down through speech rather than writing, thus making them difficult to capture using conventional methods. Quantum computing, on the other hand, could facilitate the processing required in order to analyze the large amounts of data necessary in order to encapsulate much of the languages, thus allowing linguists to create comprehensive digital archives of these endangered languages. These archives could include the transcriptions of the languages themselves, as well as the context they are placed in that are integral to the language's identity.

Quantum algorithms, such as Grover's algorithm and Shor's algorithm, can be utilized in order to enhance the efficiency of linguistic data retrieval as well as linguistic data analysis. Grover's algorithm is an algorithm that offers a quadratic speedup for databases searches that are yet to be sorted, and could therefore be used in order to quickly identify patterns specific to a language or structures within larger datasets. This capability is valuable especially in the context of language preservation, where researchers need to sort through incredibly vast recordings in order to identify uncommon linguistic features. By accelerating these processes, quantum computing is capable of significantly reducing the time and resources needed in order to document endangered languages, thus making preservation efforts much more feasible on a large scale especially.

Quantum cryptography also offers a promising avenue in order to securely transmit and store data, including in regards to linguistics. Language preservation efforts will need to increasingly rely on digital platforms, thus resulting in security and integrity within these linguistic archives in order for them to be truly effective. Quantum cryptography utilizes the principles of quantum mechanics in order to create impossible to crack encryption protocols, and could therefore provide an incredibly effective solution to this challenge. Quantum key distribution, or QKD, is a method that uses quantum states in order to securely send encryption keys, and is a protocol that could therefore be used in the transmission of sensitive linguistic data. QKD ensures that any attempt to interfere with the data is detectable, since the act of measurement in quantum mechanics by nature disturbs quantum states. Thus, QKD becomes the ideal tool for protecting endangered language archives from unwanted tampering. Thus, by employing quantum cryptographic techniques, it becomes possible and rather easy to safeguard records of endangered records, thus preserving them for observation and analysis by future generations and for future generations of the culture to which the language being recorded belongs to.

Quantum information theory is a field that explores the encoding, transmission, and manipulation of information by using quantum systems. This can also offer a perspective into how quantum principles can be applied to the field of language preservation. An area worth studying is in regards to entanglement, as was previously discussed. This non-locality thus has fascinating implications in regards to the communication and for information theory as well. Entanglement, as previously discussed, can be utilized in order to create new forms of linguistic representations.

Entanglement, for example, could be used in order to develop encoding schemes that can be used to preserve both the language itself and the context in which the language exists in, two things that are often lost in common conventional representations that use the internet, as most common representations lack the nuance, detail, and ability to preserve these crucial aspects of data preservation, resulting in imperfect preservations. Languages are not simply collections of words, symbols, and grammar rules; they are the products of the culture(s) they exist in and the context in which they are spoken. By failing to recognize that, we fail to preserve the core of the languages we must preserve for future prosperity. Quantum entanglement, with its ability to add nuance, represent, and send pertinent information, can potentially be used in order to encode these complex nuances that are coupled with linguistic data, preserving the rich culture that is integral to language.

Quantum machine learning, a subfield of quantum computing and AI that can apply quantum algorithms to machine learning tasks, allowing those machines to "learn" similarly to how a human can learn, is instrumental to the developing field of advanced linguistic models. These models can be used to reconstruct lost languages, identify and as a result preserve unique dialects, and, using their "learning" mechanism, predict the evolution of distinct languages and their dialects over time. Quantum machine learning algorithms, including quantum neural networks and quantum support vector machines, can be trained on certain linguistic datasets in order to identify patterns that are difficult, if not impossible, to detect using more classical and commonly used methods. By uncovering these patterns, machine learning can provide new insights in language diversity and thus help to develop strategies for language revitalization, thus helping to preserve language as a whole not only as a stagnant, unchanging phenomenon, but also as a key part of a culture that grows over time as the population to which it belongs grows over time.

The implications of quantum physics in regards to language preservation also extends beyond the technical, theoretical realm and into a realm that's more applicable interdisciplinarily. The interdisciplinary collaboration between quantum physicists, linguists, and cultural anthropologists can lead to new theoretical frameworks that can bridge the gap between both physical sciences as well as the humanities. Quantum physics places an emphasis on the fundamental nature of reality itself, and could therefore offer unique perspectives on the nature of language itself. For example, the probabilistic and fluid nature of quantum mechanics is capable of providing a new lens through which to understand the probabilistic and fluid nature of language itself, thus challenging the traditional, less probabilistic and fluid traditional linguistic evolution models.

In conclusion, quantum physics can be used in order to preserve language, representing a frontier that could address some of the most imminent and challenging problems in the field of physics due to the similarities between these two unique and necessary fields. Quantum computing offers the potential to both process vast amounts of data as well as to analyze it, making it perfect for the preservation of something that requires so much context and recordings in order to accurately preserve it, like language preservation. Quantum cryptography provides foolproof methods for actually transmitting linguistic data as well as its storage, as well. Both quantum information theory and quantum machine learning can lead to even more methods for the encoding, representing, and reconstruction of endangered languages, thus preserving not just the words and grammar, but also the unique culture in which the language is used. As quantum technologies continue to improve, it will become more essential for linguists and physicists to collaborate in order to ensure the preservation of endangered and extinct languages, and in order to revolutionize both fields.

Language Preservation and Accessibility

The preservation of languages requires an approach that integrates techniques from both linguistics and physics, each of which need to contribute unique ideas that address different parts of the challenge that is language preservation. Linguistics itself provides a foundation for understanding how languages are structured and how languages develop, while physics offers the methods required in order to document and analyze the data needed to actually preserve the languages. By utilizing the strengths of both physics and linguistics, we can create an interdisciplinary and comprehensive strategy for protecting endangered languages, thus ensuring that future generations have access to the unique languages that encompass the human experience as a whole.

First, we must document the languages we need to preserve thoroughly. We must do this by first collecting as much spoken data as possible directly from the native speakers of the language that requires preserving. This involves creating recordings of conversations, traditional stories, songs, and daily interactions. In some cases, linguists also document formal grammatical structures, word lists, and idiomatic expressions. These recordings and written materials form the basis of the linguistic archive. Effective preservation requires capturing the full extent of a language's features, including its phonology, or the range of sounds it uses, its morphology, or the structure of its words, its syntax, or its sentence structure, and its semantics, or the meaning of words/phrases within the languages.

Once a significant amount of data from the language that is being preserved is captured, physics is used in order to obtain the highest quality and longest lasting recordings of the language. By utilizing acoustics, we can use high-fidelity audio recordings in order to accurately capture the nuances of sound within speech, particularly in regards to a language that will have unique phonetic features not often found in languages, like the hard "r" in English found in words like "stronger" or "better". For example, some languages, like Xhosa, use clicks, while others use ejective consonants or tonal variations that are not easily expressed in written language, but are much easier to observe in recordings of languages. In order to preserve these subtleties in sound, physicists must work with linguists in order to use high-quality recording equipment in order to capture the full frequency of sounds in the language being preserved. Finally, these recordings must be digitized in formats that can preserve the original acoustic properties without lossy compression so that the language can be heard by future generations to its full extent.

Information theory, a field rooted in physics and mathematics, plays a crucial role in organizing and storing linguistic data efficiently. Information theory allows linguists to model language as a system of information transfer, which can help in determining how best to store and compress linguistic data without losing essential information. By applying principles of entropy, which measures the amount of unpredictability or information content in a message, linguists can identify which elements of the language are critical to preserve and which can be simplified without significant loss. This is particularly useful when dealing with massive archives of language recordings, as it allows for more efficient storage and retrieval while maintaining the integrity of the data.

Preserving a language also involves analyzing its structural components, and this is where physics-based computational models can assist linguists in understanding and preserving grammatical systems. Linguistic data is often complex and involves large datasets, particularly when it comes to documenting less familiar or endangered languages. By employing algorithms based on physics principles, such as wave equations or dynamical systems, computational models can simulate the evolution of a language over time, predict changes, and highlight patterns that might be invisible to the human observer. These models allow linguists to study how certain linguistic features may be more prone to loss than others in a language under pressure from more dominant languages, offering valuable insight into which aspects of the language need the most urgent preservation efforts.

In addition to recording and modeling the structure of languages, it is essential to ensure that they are accessible to future generations. One of the most effective ways to do this is to utilize the field of computational linguistics, which has an overlap between physics and linguistics themselves. By creating digital resources, like dictionaries, grammar guides and apps for the purpose of learning the language, endangered languages can become much more accessible to people who would like to learn the language but simply do not have the resources to do so. These tools utilize algorithms to analyze the language that is being portrayed in them in order to provide users with intuitive ways in which to engage with the data. For example, machine learning algorithms that are trained on large datasets of information from that language can create predictive text tools in order to give learners more examples of the language that they wish to learn. Thus, these tools can keep dead or dying languages alive by encouraging speakers to actually learn the language.

One of the major challenges in preserving languages is the risk of data loss over time. Language archives can be vast, and storing this information in physical formats like tapes or paper runs the risk of degradation or destruction. This is where the principles of quantum mechanics and modern data storage methods, rooted in physics, come into play. Quantum computing and quantum storage systems, while still in their early stages, offer the potential for far more efficient data storage with minimal risk of loss over time. By utilizing quantum bits (qubits) that can store more information than traditional binary systems, researchers are exploring ways to store enormous amounts of linguistic data in a fraction of the space required by current digital systems. This ensures that even vast repositories of language recordings can be stored securely and accessed by future generations, minimizing the risk that these languages will be forgotten.

Another key aspect in regards to language preservation is making sure the language is transmitted effectively to learners of the language. Approaches with roots in physics, such as speech synthesis and language learning software can help bridge the gap. Speech synthesis technologies allow for highly accurate digital voices to be created, such as those present in current AI technologies such as ChatGPT. These synthesized voices can be more accurate than non-native human voices in terms of replicating the way a human speaker of a language sounds, and can thus be used in educational apps, VR experiences, and in interactive language lessons if the technology develops to that point, thus providing learners with the highest-quality learning experiences possible. By using models of the acoustic properties of languages, speech synthesis can also ensure that learners know the right way to pronounce words and sounds, the rhythm, and the intonation patterns of the language, which are incredibly difficult to preserve through using written materials only.

Machine learning, another computational tool grounded in physics, also plays an increasingly important role in language preservation. By using large datasets of the language that is being preserved, machine learning algorithms can identify patterns within the language, therefore allowing us to gain new insights into how a language functions. For example, we can analyze how some sounds/grammar structures are utilized in a certain context, and machine learning algorithms can as a result predict features that are likely going to rapidly change or even die out. This information can thus be utilized to know what aspects of language to prioritize preserving, and can also simply help us know more about the language itself. More than that, machine learning algorithms can also be used to develop tutoring systems that can develop based on what the learner needs to practice, therefore making it even easier for new learners of a language to learn that language.

To complement these technical approaches, linguists also focus on sociolinguistic factors that influence language preservation. By understanding the social contexts in which languages are used, linguists can work with communities to promote language use in everyday life. Physics-based technologies, such as mobile devices and internet platforms, offer new opportunities to connect speakers of endangered languages across geographic distances. Online forums, social media platforms, and virtual meeting spaces can serve as hubs for language communities, where speakers and learners can practice the language and keep it alive. These digital environments can be enhanced by physics-based algorithms that optimize the transmission of audio and video data, ensuring that conversations remain fluid and natural even in low-bandwidth situations.

In conclusion, preserving languages requires a coordinated effort between linguistics and physics, drawing on the strengths of both disciplines to create effective strategies for documentation, analysis, and transmission. Linguists provide the theoretical framework for understanding the structure and use of languages, while physics offers the technical tools needed to capture, store, and model linguistic data. By utilizing advancements in the fields of acoustics, information theory, quantum computing, speech synthesis, and machine learning, we are capable of creating systems used for preserving endangered languages. Thus, these approaches not only ensure the preservation of languages as stagnant things that are allowed to die out in day-to-day conversation, but also continuously teach new learners how to speak these new languages, preserving the language's prosperity as living things that exist within the world's cultural memory.

Universal Languages

The idea of developing a more universally understandable form of English, optimized to be comprehensible across different languages, is an intriguing proposition that sits at the intersection of linguistics, mathematics, and physics. This concept revolves around the notion of finding patterns in phonetics, semantics, and syntax across a broad range of languages, and using these patterns to construct a version of English—or any language—that is inherently more accessible to speakers of diverse linguistic backgrounds.

One of the key aspects to consider in creating a more universally understandable English is the phonetic structure of words. Phonetics is the study of the sounds of human speech, and can offer a way in which to analyze the sounds that are most common and most easily understood as well as produce indeterminate what language someone may speak. By using principles from acoustics, which studies the properties of sound, researchers can analyze aspects of language in order to determine how to make language more accessible.

By using mathematical models and statistical analysis, linguists can analyze the phonetic structures of many distinct languages in order to identify the most common as well as the most easily produced sounds. These findings can then as a result both inform the creation of a more phonetically accessible English depending on where you are, but also increase understanding of how language itself works. For example, choosing words with phonemes that have a high cross-linguistic frequency can make this version of English easier for non-native speakers to understand with less effort.

Another aspect of creating a universally understandable form of English involves the selection of words based on their semantic universality. Lexical semantics, the study of word meanings and relationships between words, provides insights into which concepts are universally understood. By applying information theory—which deals with the quantification of information and the efficiency of communication—it is possible to analyze how effectively different words convey meaning across languages.

For example, the word "mother" is a basic, universally understood concept, but the word "mom" might not be as easily recognizable across all languages due to variations in its phonetic structure. Through a cross-linguistic frequency analysis, researchers could determine which words or roots are most commonly used to express certain concepts across a wide range of languages. By favoring these words in a universal form of English, the language could become more intuitive for speakers of other languages.

This process could involve creating a corpus (a large collection of texts) from various languages and applying machine learning algorithms to identify common lexical patterns. These patterns could then be used to standardize certain words in this new form of English, ensuring that the vocabulary is as universally comprehensible as possible.

Syntax, the arrangement of words and phrases to create well-formed sentences, is another critical area to consider. Some languages have relatively simple syntactic structures, while others are more complex, with a higher degree of inflection, word order variation, and agreement rules. A more universally understandable form of English would need to adopt a syntactic structure that is easily grasped by speakers of languages with different syntactic rules.

One approach to this is to use dynamical systems theory, a branch of mathematics used to study the behavior of complex systems over time. By modeling the evolution of syntactic structures as a dynamical system, linguists can analyze which syntactic patterns are stable and simple across different languages. This analysis could reveal a syntactic "core" that is less likely to cause confusion for non-native speakers.

For example, languages that follow a Subject-Verb-Object (SVO) word order, such as English, Mandarin, and Spanish, might form the basis of a universal syntax because this word order is relatively common and straightforward. Simplifying the syntax to minimize the use of complex clauses and reducing exceptions to grammatical rules could also make the language easier to learn and use universally.

Another important consideration is cognitive load—the amount of mental effort required to process information. Physics-based models of cognitive processes, particularly those used in neuroscience, can help predict how easily different linguistic structures are processed by the brain. By analyzing how quickly and accurately people from different linguistic backgrounds process certain words, sounds, and syntactic structures, it is possible to design a version of English that minimizes cognitive load and maximizes comprehension.

For instance, reaction time experiments could be conducted to measure how quickly speakers of different languages recognize and understand various English words and phrases. By selecting words and structures that consistently result in faster processing times across diverse linguistic groups, a more universally accessible form of English could be developed.

Finally, entropy—a concept from information theory that measures uncertainty or unpredictability—can be applied to language to ensure that the universal form of English maintains an optimal balance between predictability and expressiveness. In communication, a certain level of redundancy (repetition of information) is useful because it ensures that messages are understood even in noisy or ambiguous situations. However, too much redundancy can make communication inefficient.

By analyzing the entropy of different linguistic elements across languages, researchers can identify which structures and word choices provide the optimal level of redundancy for universal comprehension. This might involve using statistical models to determine how different languages handle redundancy and adjusting the new form of English to include similar levels of repetition, ensuring that the language is both clear and efficient.

The development of a more universally accessible form of English can be achieved by utilizing aspects of phonetics, lexical semantics, syntax, cognitive science, information theory, and more linguistic and auditory processes. This is all underpinned by a physical analysis in order to interpret resulting data and form this "new" form of English. This approach not only leverages the universal aspects of human language but also applies scientific principles to enhance communication in an increasingly interconnected world.

Emojis are another realm for universal communication. The idea that emojis can, and are, evolving into a common language for communication among speakers of different languages has quite a bit of potential, and is already happening to some extent. Because emojis serve as a visual language based on symbols for emotions that are, for the most part, universal irrespective of language, results in people having the ability to communicate with greater ease and efficiency. More than that, even among speakers of the same language but of different cultures can utilize emojis in order to communicate with more ease; while those from Australian who say "thong" are talking about sandals, those from the USA who say "thong" mean underwear, resulting in a miscommunication that can be resolved by the visual representation of a sandal or the context in which it lies. The potential for emojis to become a universal way of communicating lies in its simplicity, as well as their expressiveness, their ease of use, and the fact that they rely on universal human experiences.

Emojis are designed to represent inherent human emotions, actions, objects, and symbols. For example, a sandal will always be a sandal, irrespective of culture. An angry-looking face and a smile will always represent the same thing, irrespective of culture or language as well. This is due to the fact that emojis represent complex, often difficult to express emotions and symbols into things that are easily recognizable and representable. The fact that emojis are pictures is also reminiscent of ancient writing systems, similar to hieroglyphics, which also relied on symbols for understanding.

From a linguistic perspective, emojis serve as a modern day form of ideograms, which is a form of language where each symbol contains a meaning that can be interpreted regardless of native language. Semiotics is the study of signs and their usage, and can provide a theoretical framework through which to understand how emojis function in communication. In semiotic terms, each emoji serves as a signifier, or a form, that can represent a signified, or the concept it conveys. A realm for ambiguity in terms of emojis is the unique meaning that emojis can contain based on their cultural context. For example, for younger generations, the crying-laughing emoji used to mean that you find something so funny it is making you cry-laugh. However, over time, it has developed into being sarcastic; it is most commonly used when something is not funny, and someone is trying to express that and say they find it funny sarcastically. While it is still used to represent crying-laughing, this new usage is worth noting. Similarly, the crying emoji, where the emoji is sobbing, now represents finding something funny or as inciting some type of emotion, instead of simply sadness.

Thus, while emojis can be, and are, used to express ideas irrespective of native language, it is also important to note that in some instances, emojis represent ideas based on the culture that they lie within.

Emojis are powerful as a universal language because of their ability to convey complex emotions as well as contextual nuances with less ambiguity. For example, a single emoji can communicate laughter, enjoyment, and genuine joy simultaneously without attempting to ambiguously explain that that is what you are feeling, making it a flexible and useful tool in the realms of digital communication. Thus, the versatility in emojis, as one emoji can also express multiple ideas when coupled with another emoji or based on the context in which it lies, can also contribute to its effectiveness as a tool that can be used to communicate irrespective of native language.

Information theory, particularly the concept of entropy, can be used to analyze how efficiently emojis convey information across different languages. In this context, entropy refers to the unpredictability or uncertainty in a message. Emojis tend to lower the entropy of a message by providing clear, visual cues that reduce the ambiguity of the text. This reduction in ambiguity is particularly valuable in cross-linguistic communication, where textual nuances might be lost or misunderstood.

By standardizing the use of certain emojis to represent specific ideas or emotions, it is possible to create a more consistent and universally understood set of symbols. This process would involve analyzing which emojis are most frequently and consistently used across different cultures and languages, and then promoting these as the core symbols in a global emoji language.

The human brain is designed to process visual information effectively as well as quickly, which is one important reason as to why emojis are so effective. Cognitive science and neuroscience can provide insights into why and how emojis are capable of resonating so well cross culturally when utilized correctly. The brain's visual cortex is designed to recognize visual symbols and place them in context, especially those that represent things designed to represent the human experience. This visual processing is faster, for the most part, than linguistic processing, especially when doing so in your non-native language, as it involves multiple cognitive steps.

In practical terms, this means that when people use emojis, they are tapping into a form of communication that the brain is naturally optimized to handle. This could explain why emojis are so popular in digital communication—they provide a quick and intuitive way to convey complex emotions and ideas that might take longer to express in words.

In order for emojis to develop into a common language, it is necessary for there to be a standardization process similar to the ones that govern natural languages. The Unicode Consortium is something that manages the standardization of emojis, and thus plays a crucial role in this regard. By continuously expanding the emoji set to include more and more symbols, the Consortium is thus helping to create a more comprehensive emoji language that is able to encompass more and more universal experiences.

While emojis are powerful, they do have limitations. Because of how context dependent they are, their meaning can vary significantly based on cultural background, social context, and textual context, as has been discussed earlier. For example, the usage of the "folded hands" emoji could represent praying in some contexts and cultures, and could also represent a thank you or a greeting in other cultures, and, finally, could represent a gesture of respect in other cultures and contexts. Thus, these variations serve as a challenge in the development of a universal emoji language, making a truly universal emoji language difficult, if not downright impossible in general.

Moreover, emojis primarily convey emotions, simple actions, and concrete objects. They are less effective at conveying abstract concepts, detailed information, or complex narratives. To address this limitation, future developments might include the creation of new, more abstract emojis or the use of combinations of emojis to express more complex ideas—a practice that is already emerging in some digital communication.

The potential for emojis to evolve into a common language for people who speak different languages is significant, particularly as digital communication continues to play a central role in global interactions. Emojis leverage the brain's natural affinity for visual symbols, reduce the entropy of communication, and offer a universal means of expressing basic human emotions and actions. While challenges remain in terms of cultural differences and the limitations of emoji expressiveness, the ongoing standardization efforts by organizations like the Unicode Consortium, coupled with advances in AI, suggest that emojis could indeed become a more integral part of global communication. As this evolution continues, emojis may bridge linguistic divides, offering a simple yet powerful tool for fostering understanding across diverse cultures.

Conclusion

The convergence of physics and linguistics in the effort to preserve endangered languages represents an interdisciplinary approach that brings together the scientific rigor of physical principles and the cultural sensitivity of linguistic studies. As discussed within this text, physics, as it is focused on laws that govern the world as a whole, offers a large variety of tools as well as methods through which we can help preserve languages. While the technical side of language preservation is incredibly necessary for the process, it is important to remember that the heart of language preservation is the people who actually speak the dying languages we must work to preserve. Thus, the role of physics in this context is as a helper through which to uplift these people and through which to pass down the vast array of knowledge that these dying languages contain.

A central theme in this book has been how we can use acoustics in order to record, analyze, and observe the sounds within different languages. The understanding of these acoustics allows us to capture the sounds of different languages with precision, thus ensuring that the subtle nuances within these sounds are preserved. Through high-fidelity recordings, we are capable of creating archives that hold not just the written symbols of a language but its living, breathing form. These archives, powered by advancements in digital storage, provide a crucial repository of linguistic data, available for future researchers, educators, and language revitalization efforts.

The application of physics in this domain does not stop at mere recording. Sophisticated analysis tools, such as spectrograms and Fourier transforms, allow us to break down speech into its component frequencies, giving us an in-depth view of how languages function on a sonic level. For linguists, this kind of analysis is invaluable, as it allows for the detailed examination of phonetic structures that may be entirely unfamiliar in widely spoken languages. For example, languages with click consonants or tonal systems that use pitch variation to change meaning present particular challenges in preservation. The ability to accurately analyze and reproduce these sounds is critical for language learners, who must grasp not only the grammar and vocabulary but also the correct pronunciation to truly bring a language back to life. This is where the partnership between physics and linguistics flourishes, as the precision of acoustic analysis informs the broader linguistic goals of preserving meaning and structure.

However, analyzing language is simply the first step in the arduous, challenging process of language preservation. One of the reasons this process is so challenging is because it is incredibly difficult to store such a vast amount of information while simultaneously maintaining its accessibility for future generations. The principles essential in information theory that are rooted in the work of pioneers such as Claude Shannon provide a framework through which we can analyze linguistic data to the fullest extent of our abilities. By applying physical concepts such as entropy, redundancy, and error correction, we are also capable of ensuring that such a vast amount of data can be stored for posterity. As digital storage technologies continue to evolve, including promising developments in quantum computing and data compression, we have the potential to safeguard massive amounts of linguistic data for centuries, if not indefinitely.

Yet, storage alone is not enough. A language that is preserved but inaccessible is effectively lost. Here, the principles of communication physics become crucial, as they enable the transmission of linguistic data across vast distances and time periods. We are fortunate to live in an age in which we have high-speed internet, cloud computing, and other technologies that make it easier than ever before in the history of humanity to connect speakers of endangered languages with those who wish to learn the endangered language, irrespective of location. Technological advancements further ensure that even the most remote communities are capable of preserving their languages.

Another important aspect of language preservation is the modeling and simulation of language systems. This is where the computational tools derived from physics can be most useful. Languages, like many systems studied in physics, exhibit complex, dynamic behavior. They evolve over time, influenced by social, cultural, and environmental factors. By using computational models to simulate these dynamics, we can gain insights into how languages change, what features are most likely to be lost, and how external pressures—such as the influence of dominant languages—affect smaller linguistic communities. These models, grounded in physical principles such as chaos theory and dynamical systems, offer a powerful way to predict and intervene in the process of language extinction, guiding preservation efforts toward the most vulnerable aspects of a language.

Speech synthesis, an area of research that blends physics, computer science, and linguistics, also offers exciting possibilities for language preservation. By using algorithms that generate speech from textual input, we can create digital voices that faithfully reproduce the sounds of endangered languages. These synthetic voices, powered by advances in waveform generation and machine learning, allow for the creation of language-learning tools, interactive educational programs, and even virtual environments where speakers can practice their language in immersive settings. The ability to synthesize speech accurately not only aids in language revitalization but also ensures that the auditory experience of a language is never truly lost, even if native speakers are no longer available to teach it.

The role of machine learning and artificial intelligence in language preservation cannot be understated. These technologies, which draw heavily on statistical physics and algorithmic principles, allow for the analysis of massive linguistic datasets, uncovering patterns and structures that might be invisible to the human eye. For example, AI algorithms can be trained to analyze spoken language and generate predictive models that help linguists identify the most important features of a language for preservation. By automating certain aspects of linguistic analysis, machine learning frees up human researchers to focus on more nuanced, interpretative work, creating a partnership between human expertise and machine efficiency that enhances the overall preservation effort.

While the technical, physical side of language preservation is incredibly essential, we must remember the heart of the goal of language preservation; the languages themselves. Languages exist as living, dynamic entities, and must be treated as such. A language that is recorded for future generations but is no longer spoken, or no longer spoken in the way that it is recorded is not a language that is truly alive. Thus, we must keep people as our focus in language preservation. Physics can lay the groundwork for language preservation, but it is the people who actually speak these languages, and the people who wish to speak these languages, that actually are necessary for true language preservation and to truly keep these languages alive. The success of language preservation hinges on the creation of environments that encourage humans to learn these languages with as much ease as is possible.

In order to accomplish this goal, we must create links between physicists, linguists, and the speakers and learners of these languages. We must involve the people at the heart of this issue in order to effectively tackle this issue. By combining the deep cultural knowledge of the speakers of the languages, and the expertise of the professionals working in these realms, we can create societies that foster the learning of these languages that are dying. Digital resources, for example, should be designed with explicit input from native speakers, and not just from physicists and linguists who only have an academic understanding of the language. Context is necessary for true understanding.

The effort to preserve languages mirrors the intrinsically human endeavor that is the desire to pass down our knowledge and culture to future generations. We work to conserve the art, culture, and history of the past, and similarly, we must work to conserve the languages that carry on the knowledge and culture of the world. Physics is necessary in order to actually accomplish this goal. Linguistics, meanwhile, provides the cultural and intellectual framework that ensures we understand the meaning and significance of the languages we seek to preserve, rendering them meaningful.

As we move forward in this interdisciplinary effort, language preservation looks to be increasingly possible, effective, and efficient. As the technology behind language preservation advances, new opportunities in regards to how we record as well as how we transmit linguistic knowledge open up. Simultaneously, we are becoming increasingly aware of the importance of cultural diversity and therefore our linguistic heritages. By combining physics with the insights we gain through linguistics, we are given the tools necessary to meet the challenge of physics head on.

Finally, in conclusion, the preservation of languages is not just a technical task we have a set protocol in order to accomplish; it's a deeply human task with real people, often deeply tied emotionally to the task, at the center of it. Physics provides us with the means to preserve language, but it is our connections with others that actually fuel our desire to preserve language. Every language represents each individual that speaks it, and thus each language is worth preserving. As we continue to interact with the world, and the worlds of others, it is important to remember this: we may differ based on language, but we are all human, and all of our backgrounds are meaningful and worth preservation.

Sources
References

References

Chandler, D. (2007). Semiotics: The Basics (2nd ed.). Routledge.

MacKay, D. J. C. (2003). Information Theory, Inference, and Learning Algorithms. Cambridge University Press.

Kuhl, P. K. (2004). Early language acquisition: Cracking the speech code. Nature Reviews Neuroscience, 5(11), 831-843.

Schmandt-Besserat, D. (1996). How Writing Came About. University of Texas Press.

Unicode Consortium. (n.d.). Unicode Standard, Version 14.0. Retrieved from https://www.unicode.org/standard/standard.html

References

Chomsky, N. (1957). Syntactic Structures. Mouton.

Ladefoged, P., & Johnson, K. (2014). A Course in Phonetics (7th ed.). Cengage Learning.

Shannon, C. E., & Weaver, W. (1949). The Mathematical Theory of Communication. University of Illinois Press.

MacKay, D. J. C. (2003). Information Theory, Inference, and Learning Algorithms. Cambridge University Press.

Kuhl, P. K. (2004). Early language acquisition: Cracking the speech code. Nature Reviews Neuroscience, 5(11), 831-843.

References

Lieberman, E., Michel, J.-B., Jackson, J., Tang, T., & Nowak, M. A. (2007). Quantifying the evolutionary dynamics of language. Nature, 449(7163), 713-716.

Nowak, M. A., Plotkin, J. B., & Krakauer, D. C. (1999). The evolution of syntactic communication. Nature, 404(6777), 495-498.

Shannon, C. E. (1948). A mathematical theory of communication. Bell System Technical Journal, 27(3), 379-423.

Zipf, G. K. (1949). Human Behavior and the Principle of Least Effort. Addison-Wesley Press.

Barabási, A.-L. (2002). Linked: The New Science of Networks. Perseus Publishing.

References

Chandler, D. (2007). Semiotics: The Basics (2nd ed.). Routledge.

Shannon, C. E., & Weaver, W. (1949). The Mathematical Theory of Communication. University of Illinois Press.

MacKay, D. J. C. (2003). Information Theory, Inference, and Learning Algorithms. Cambridge University Press.

Pereira, F. C. N., & Grosz, B. J. (1994). Natural language processing and its role in artificial intelligence. Artificial Intelligence, 63(1-2), 1-64.

Sperber, D., & Wilson, D. (1995). Relevance: Communication and Cognition (2nd ed.). Blackwell.

References

Fant, G. (1970). Acoustic Theory of Speech Production: With Calculations Based on X-Ray Studies of Russian Articulations. Mouton.

Ladefoged, P., & Johnson, K. (2014). A Course in Phonetics (7th ed.). Cengage Learning.

Johnson, K. (2011). Acoustic and Auditory Phonetics (3rd ed.). Wiley-Blackwell.

Boersma, P., & Weenink, D. (2020). Praat: Doing Phonetics by Computer [Computer program]. Version 6.1.16, retrieved from http://www.praat.org/.

Xu, Y. (2005). Speech melody as articulatorily implemented communicative functions. Speech Communication, 46(3-4), 220-251.

https://www.lifeprint.com/asl101/pages-layout/grammar.htm
https://www.iesalc.unesco.org/en/2022/02/21/a-decade-to-prevent-the-disappearance-of-3000-languages/
https://www.ncbi.nlm.nih.gov/pmc/articles/PMC4321236/
https://storiesfromtanya.com/2015/09/18/a-whole-lot-of-shi/
https://www.ncbi.nlm.nih.gov/pmc/articles/PMC5529419/

https://www.researchgate.net/figure/Fig-S2-Heat-map-based-on-density-of-languages-with-and-without-complex-tonality-3_fig3_271223871
https://academicjournals.org/article/article1379500755_Shakib.pdf
https://mural.maynoothuniversity.ie/12890/1/Stifter%20Encyc.pdf
https://udaras.ie/en/our-language-the-gaeltacht/history-of-the-irish-language/
https://www.merriam-webster.com/wordplay/norman-conquest-new-english-words
https://www.ncbi.nlm.nih.gov/pmc/articles/PMC4176592/
https://commons.wikimedia.org/wiki/File:Zipf%27s_law_on_War_and_Peace.png
https://encyclopedia.pub/entry/34576

References
Fant, G. (1970). Acoustic Theory of Speech Production: With Calculations Based on X-Ray Studies of Russian Articulations. Mouton.

Ladefoged, P., & Johnson, K. (2014). A Course in Phonetics (7th ed.). Cengage Learning.

Johnson, K. (2011). Acoustic and Auditory Phonetics (3rd ed.). Wiley-Blackwell.

Boersma, P., & Weenink, D. (2020). Praat: Doing Phonetics by Computer [Computer program]. Version 6.1.16, retrieved from http://www.praat.org/.

References
1. Jakobson, R., Fant, G., & Halle, M. (1952). Preliminaries to Speech Analysis: The Distinctive Features and their Correlates. MIT Press.
2. Trubetzkoy, N. S. (1939). Grundzüge der Phonologie. Vandenhoeck & Ruprecht.
3. Ladefoged, P., & Johnson, K. (2014). A Course in Phonetics (7th ed.). Cengage Learning.
4. Kent, R. D., & Read, C. (2002). The Acoustic Analysis of Speech. Singular Publishing Group.

5. Fitch, W. T. (2010). The Evolution of Language. Cambridge University Press.
6. Lieberman, P. (1984). The Biology and Evolution of Language. Harvard University Press.
Xu, Y. (2005). Speech melody as articulatorily implemented communicative functions. Speech Communication, 46(3-4), 220-251.

References
1. Jakobson, R., Fant, G., & Halle, M. (1952). Preliminaries to Speech Analysis: The Distinctive Features and their Correlates. MIT Press.
2. Trubetzkoy, N. S. (1939). Grundzüge der Phonologie. Vandenhoeck & Ruprecht.
3. Ladefoged, P., & Johnson, K. (2014). A Course in Phonetics (7th ed.). Cengage Learning.
4. Kent, R. D., & Read, C. (2002). The Acoustic Analysis of Speech. Singular Publishing Group.
5. Fitch, W. T. (2010). The Evolution of Language. Cambridge University Press.
6. Lieberman, P. (1984). The Biology and Evolution of Language. Harvard University Press.

References
1. Busemeyer, J. R., & Bruza, P. D. (2012). *Quantum Models of Cognition and Decision*. Cambridge University Press.

References
Whorf, B. L. (1956). Language, Thought, and Reality: Selected Writings of Benjamin Lee Whorf. MIT Press.
Sapir, E. (1929). The Status of Linguistics as a Science. Language, 5(4), 207-214.
Gell-Mann, M. (1994). The Quark and the Jaguar: Adventures in the Simple and the Complex. W.H. Freeman and Company.
2. Piotrowski, K. (2019). Quantum Linguistics: Theoretical Applications of Quantum Mechanics in Linguistic Research. *International Journal of Applied Linguistics & English Literature*, 8(2), 1-8.

3. Widdows, D. (2004). *Geometry and Meaning*. CSLI Publications.
4. Aerts, D., & Aerts, S. (1995). Applications of Quantum Statistics in Psychological Studies of Decision-Processes. *Foundations of Science*, 1(1), 85-97.
5. Hameroff, S., & Penrose, R. (2014). Consciousness in the Universe: A Review of the 'Orch OR' Theory. *Physics of Life Reviews*, 11(1), 39-78

References

Nielsen, M. A., & Chuang, I. L. (2010). *Quantum Computation and Quantum Information: 10th Anniversary Edition*. Cambridge University Press.

Shor, P. W. (1997). Polynomial-Time Algorithms for Prime Factorization and Discrete Logarithms on a Quantum Computer. *SIAM Journal on Computing*, 26(5), 1484-1509.

Grover, L. K. (1996). A Fast Quantum Mechanical Algorithm for Database Search. *Proceedings of the Twenty-Eighth Annual ACM Symposium on Theory of Computing*, 212-219.

Gisin, N., Ribordy, G., Tittel, W., & Zbinden, H. (2002). Quantum Cryptography. *Reviews of Modern Physics*, 74(1), 145-195.

Biamonte, J., Wittek, P., Pancotti, N., Rebentrost, P., Wiebe, N., & Lloyd, S. (2017). Quantum Machine Learning. *Nature*, 549(7671), 195-202.

http://hyperphysics.phy-astr.gsu.edu/hbase/Waves/funhar.html

www.ingramcontent.com/pod-product-compliance
Lightning Source LLC
Chambersburg PA
CBHW060618200326
41521CB00007B/802